1st International Workshop on Theoretical and Computational Physics: Condensed Matter, Soft Matter and Materials Physics & 38th National Conference on Theoretical Physics 2013

Journal of Physics: Conference Series Volume 537

Da Nang, Vietnam
29 July – 1 August 2013

ISBN: 978-1-63439-372-0
ISSN: 1742-6588

Printed from e-media with permission by:

Curran Associates, Inc.
57 Morehouse Lane
Red Hook, NY 12571

Some format issues inherent in the e-media version may also appear in this print version.

Printed by Curran Associates, Inc. (2014)

For permission requests, please contact the Institute of Physics at the address below.

Institute of Physics
Dirac House, Temple Back
Bristol BS1 6BE UK

Phone: 44 1 17 929 7481
Fax: 44 1 17 920 0979

techtracking@iop.org

Additional copies of this publication are available from:

Curran Associates, Inc.
57 Morehouse Lane
Red Hook, NY 12571 USA
Phone: 845-758-0400
Fax: 845-758-2634
Email: curran@proceedings.com
Web: www.proceedings.com

TABLE OF CONTENTS

1ST INDUSTRIAL WORKSHOP ON THEORETICAL AND COMPUTATIONAL PHYSICS (IWTCP1): CONDENSED MATTER, SOFT MATTER AND MATERIALS PHYSICS

38TH NATIONAL CONFERENCE ON THEORETICAL PHYSICS (NCTP38)

Journal of Physics: Conference Series **537** (2014) 011001 doi:10.1088/1742-6596/537/1/011001

Preface

This volume contains selected papers presented at the 38th National Conference on Theoretical Physics (NCTP-38) and the 1st International Workshop on Theoretical and Computational Physics: Condensed Matter, Soft Matter and Materials Physics (IWTCP-1). Both the conference and the workshop were held from 29 July to 1 August 2013 in Pullman hotel, Da Nang, Vietnam.

The IWTCP-1 was a new activity of the Vietnamese Theoretical Physics Society (VTPS) organized in association with the 38th National Conference on Theoretical Physics (NCTP-38), the most well–known annual scientific forum dedicated to the dissemination of the latest development in the field of theoretical physics within the country. The IWTCP-1 was also an External Activity of the Asia Pacific Center for Theoretical Physics (APCTP).

The overriding goal of the IWTCP is to provide an international forum for scientists and engineers from academia to share ideas, problems and solution relating to the recent advances in theoretical physics as well as in computational physics. The main IWTCP motivation is to foster scientific exchanges between the Vietnamese theoretical and computational physics community and world–wide scientists as well as to promote high–standard level of research and education activities for young physicists in the country.

About 110 participants coming from 10 countries participated in the conference and the workshop. 4 invited talks, 18 oral contributions and 46 posters were presented at the conference. In the workshop we had one keynote lecture and 9 invited talks presented by international experts in the fields of theoretical and computational physics, together with 14 oral and 33 poster contributions.

The proceedings were edited by Nguyen Tri Lan, Trinh Xuan Hoang, and Nguyen Ai Viet.

We would like to thank all invited speakers, participants and sponsors for making the conference and the workshop successful.

Nguyen Ai Viet Chair of NCTP-38 and IWTCP-1

Journal of Physics: Conference Series **537** (2014) 011001 doi:10.1088/1742-6596/537/1/011001

Organizing Institutions

- Institute of Physics, Vietnam Academy of Science and Technology (IOP-VAST)
- University of Science, Vietnam National University, Hanoi (HUS-VNU)
- Vietnam National University, Ho Chi Minh City (VNU HCM)

Sponsors

- National Foundation for Science and Technology Development (NAFOSTED)
- Asia Pacific Center for Theoretical Physics (APCTP)
- Institute of Physics, Vietnam Academy of Science and Technology (IOP-VAST)
- University of Science, Vietnam National University, Hanoi (HUS-VNU)
- Vietnam National University, Ho Chi Minh city (VNU-HCM)
- Vietnam Academy of Science and Technology (VAST)
- Ho Chi Minh City University of Pedagogy (HCMUP)

Journal of Physics: Conference Series **537** (2014) 011001 doi:10.1088/1742-6596/537/1/011001

Committees

NCTP-38	IWTCP-1
	Honorary Chairmen Nguyen Van Hieu (VAST) Tran Thanh Van (Rencontres du Vietnam) Peter Fulde (APCTP)
Conference Chairman Nguyen Ai Viet (IOP-VAST), Chair	**Workshop Chairmen** Nguyen Ai Viet (IOP-VAST), Chair Nguyen Manh Duc (CCFE, UKAEA), Co-chair
	International Advisory Board Paolo Carloni (GRSSS, Julich) Tae-Soo Chon (Pusan National University) Hung T. Diep (University of Cergy-Pontoise) Nguyen Manh Duc (CCFE, UKAEA) Yuan Ping Feng (NUS, Singapore) Peter Fulde (APCTP) Jisoon Ihm (Seoul National University) Hideaki Kasai (Osaka University) Yoshiyuki Kawazoe (Tohoku University) Mai Suan Li (Polish Academy of Sciences) Nguyen Van Lien (IOP-VAST) Dragica Vasileska (Arizona State University) Nguyen Ai Viet (IOP-VAST)
Organizing Committee Trinh Xuan Hoang (IOP-VAST), Chair Bach Thanh Cong (HUS, VNU) Hoang Dzung (VNU HCM) Nguyen Tri Lan (IOP-VAST) Do Van Nam (HUST) Nguyen Hong Quang (VAST) Hoang Anh Tuan (IOP-VAST)	**Organizing Committee** Trinh Xuan Hoang (IOP-VAST), Chair Bach Thanh Cong (HUS, VNU) Do Tien Dzung (NAFOSTED) Hoang Dzung (VNU HCM) Nguyen Tri Lan (IOP-VAST) Do Van Nam (HUST) Nguyen Hong Quang (VAST) Hoang Anh Tuan (IOP-VAST)
Program Committee Nguyen Toan Thang (IOP-VAST), Chair Bach Thanh Cong (HUS, VNU) Ho Trung Dung (IoP HCM, VAST) Phung Van Dong (IOP-VAST) Vu Ngoc Tuoc (HUST)	**Program Committee** Nguyen Toan Thang (IOP-VAST), Chair Bach Thanh Cong (HUS, VNU) Ho Trung Dung (IOP HCM, VAST) Vu Ngoc Tuoc (HUST)
Proceedings Nguyen Tri Lan (IOP-VAST)	**Proceedings** Nguyen Tri Lan (IOP-VAST)
Secretariat Chu Thuy Anh (IOP-VAST)	**Secretariat** Chu Thuy Anh (IOP-VAST)

Journal of Physics: Conference Series **537** (2014) 011001 doi:10.1088/1742-6596/537/1/011001

List of Participants

ANH-TUAN HOANG

Institute of Physics, VAST

Email: hatuan@iop.vast.ac.vn

BUI THI LE QUYEN

Đại học Y Thái Bình

Email: builequyenytb@gmail.com

BACH THANH CONG

University of Science, VNU, Hanoi

Email: congbt@vnu.edu.vn

BUI DUC HUNG

University of Science, VNU, Hanoi

Email: hungphysics.qn@gmail.com

BUI DUC TINH

Hanoi National University of Education

Email: tinhbd@hnue.edu.vn

BUI PHUONG THUY

Nam Dinh University of Technology Education

Email: thuy_24784@yahoo.com

BÙI THANH LIÊM

Institute of Physics, Vietnam Academy of Science and Technology

Email: nhviet@iop.vast.ac.vn

BUI VAN PHO

Graduate School of Engineering, Osaka University

Email: buivanpho@up.prec.eng.osaka-u.ac.jp

CAO HUY PHUONG

Hung Vuong University, Phu Tho

Email: caohuyphuongvlpt@yahoo.com.vn

CAO THI BICH

Institute of Physics, VAST

Email: ctbich@iop.vast.ac.vn

CHIA-CHU CHEN

National Cheng-Kung University, Physics Department

Email: chiachu@phys.ncku.edu.tw

CHUONG HA-HUNG NGUYEN

University of Science at Ho Chi Minh City

Email: nhhchuong@phys.hcmuns.edu.vn

DUY QUANG TO

VNU University of Science

Email: quang.toduy@gmail.com

ĐẶNG VĂN SOA

Hanoi National University of Education

Email: dvsoa@hnue.edu.vn

DAO THI THUY NGA

Trường Trung học phổ thông Hàn Thuyên

Email: daothuyngaht@gmail.com

DAO VONG DUC

Institute of Physics, VAST

Email: dvduc@iop.vast.ac.vn

NGUYEN VAN THANH

Khoa Vật Lý, Trường Đại Học Khoa Học Tự Nhiên

Email: thanhpc1987@gmail.com

DO NGOC SON

HoChiMinh City University of Technology

Email: dns.vn2013@yahoo.com

DUONG DAI PHUONG

Đoàn 871 Bộ Quốc Phòng

Email: vanha318@yahoo.com

DUY KHUONG PHUNG

Institute of Physics, VAST

Email: pdkhuong@iop.vast.ac.vn

FLAVIO SENO

Dipartimento di Fisica e Astronomia, Universita' di Padova

Email: flavio.seno@pd.infn.it

HUNG THE DIEP

Laboratoire de Physique Théorique et Modélisation, UMR 8089, University of Cergy-Pontoise, CNRS, France

Email: diep@u-cergy.fr

HA THANH HUNG

Hanoi National University of Education No. 2

Email: hthung@grad.iop.vast.ac.vn

HA VIET ANH

Hanoi University of Science and Technology

Email: haviet.anh52@gmail.com

HIDEAKI KASAI

Department of Applied Physics, Osaka University

Email: kasai@dyn.ap.eng.osaka-u.ac.jp

HIEU NGUYEN T.

Institute of Computational Science and Technology, Ho Chi Minh City

Email: nthieu.phys@gmail.com

HO KHAC HIEU

Trường Đại học Xây dựng

Email: hieuhk@nuce.edu.vn

HO SI TA

Department of Physics, National University of Civil Engineering.

Email: hositadhxd@gmail.com

HOANG DUNG

University of Science, National University of Hochiminh City

Email: hdung@vnuhcm.edu.vn

HOANG NGOC LONG

Institute of Physics, VAST

Email: hnlong@iop.vast.ac.vn

HOANG VAN HUNG

Ho Chi Minh City University of Pedagogy

Email: hunghv@hcmup.edu.vn

HOANG-DO NGOC-TRAM

Ho Chi Minh City University of Pedagogy

Email: ngoctramhd@gmail.com

HONG-SON NGUYEN

Trường Đại học Công đoàn

Email: atvsld@gmail.com

HUY-VIET NGUYEN

Institute of Physics, Vietnam Academy of Science and Technology

Email: nhviet@iop.vast.ac.vn

HUỲNH ANH HUY

Bộ môn Sư phạm Vật lý - Khoa Sư phạm - ĐH Cần Thơ

Email: hahuy@ctu.edu.vn

HUYNH VINH PHUC

Dong Thap University

Email: hvphuc@dthu.edu.vn

JONG HYUN JUNG

Seoul National University

Email: jjhjm17@snu.ac.kr

KI-SEOK KIM

POSTECH

Email: tkfkd@postech.ac.kr

LE THO HUE

Institute of Physics, VAST

Email: lthue@iop.vast.ac.vn

LE BIN HO

Ho Chi Minh City Institute of Physics

Email: bin262@gmail.com

LE HOANG ANH

Hanoi University of Science and Technology

Email: anh.lehoang@hust.edu.vn

LE MINH THU

Department of Physics, Hanoi National University of Education

Email: minhthule1981@gmail.com

LE THI CAM TU

Ho Chi Minh City University of Pedagogy

Email: letu106@gmail.com

LE TUAN

Hanoi University of Science and Technology

Email: letuan_2002@yahoo.com

LÊ VĂN HOÀNG

Ho Chi Minh City University of Pedagogy

Email: hoanglg@hcmup.edu.vn

LƯU THỊ KIM THANH

Hanoi National University of Education No. 2

Email: luuthikim.thanh888@gmail.com

MAI CHUNG NGUYEN

Institute of Physics, VAST

Email: nmchung@iop.vast.ac.vn

MAI-SUAN LI

Instiute for Computational Science and Technology at Ho Chi Minh City, Quang Trung Software City, Tan Chanh Hiep Ward, District 12, HCM city, Vietnam

Email: masli@ifpan.edu.pl

MINSUNG KIM

Department of Physics and Astronomy, Seoul National University

Email: eriof00@snu.ac.kr

NGUYEN HUY THAO

Hanoi National University of Education No. 2

Email: nhthao@grad.iop.vast.ac.vn

NGUYEN AI VIET

Information Technology Institute, Vietnam National University, Hanoi

Email: naviet@vnu.edu.vn

NGUYEN AI VIET

Institute of Physics, VAST

Email: vieta@iop.vast.ac.vn

NGUYEN BA HUNG

Bộ môn Vật lý - Lý sinh, Học viện Quân Y

Email: hungnb80@yahoo.com.vn

NGUYỄN CHÍNH CƯƠNG

Hanoi National University of Education

Email: cuongnc@hnue.edu.vn

NGUYEN DANH TUNG

Institute of Physics, VAST

Email: ndtung@iop.vast.ac.vn

NGUYEN DUC HIEN

Chư Păh - Gia Lai

Email: n.duchien@gmail.com

NGUYEN HUU VINH

Ho Chi Minh City University of Technology

Email: vinh_mytho@yahoo.com

NGUYEN MANH DUC

Culham Center for Fusion Energy, UKAEA

Email: duc.nguyen@ccfe.ac.uk

NGUYEN NGOC HIEU

Center of Research and Development, Duy Tan University

Email: hieunguyenvly@yahoo.com

NGUYỄN NGỌC TY

Ho Chi Minh City University of Pedagogy

Email: nguyenngocty1182@gmail.com

NGUYEN NHU LE

Khoa lý, Trường Đại học Sư phạm Huế

Email: naspears13@gmail.com

NGUYỄN PHƯƠNG DUY ANH

Trường Đại học Thủ Dầu Một

Email: nguyenphuongduyanh@yahoo.com

NGUYEN QUANG BAU

Faculty of Physics, Hanoi University of Science, Vietnam National University

Email: nguyenquangbau54@gmail.com

NGUYEN QUANG HUNG

Tan Tao University

Email: hung.nguyen@ttu.edu.vn

NGUYEN QUOC THAI

Dong Thap University

Email: nqthai@dthu.edu.vn

NGUYEN THANH PHONG

Can Tho University

Email: thanhphong@ctu.edu.vn

NGUYỄN THÀNH SƠN

Đại học Kiến Trúc Tp. Hồ Chí Minh

Email: thanhsonsply@gmail.com

NGUYEN THANH TIEN

College of Science, Can Tho University

Email: nttien@ctu.edu.vn

NGUYEN THI HA LOAN

Hanoi National University of Education No. 2

Email: hloansp2@gmail.com

NGUYEN THI KIEU THU

Khoa Vật lý và Vật lý kỹ thuật, Đại học Khoa học tự nhiên TP HCM

Email: kieuthu2212@gmail.com

NGUYEN THI LE THUY

Hue University

Email: t.l.thuy.nguyen@gmail.com

NGUYEN THI PHUONG LAN

Hanoi National University of Education No. 2

Email: lanntphuong@gmail.com

NGUYEN THI THU NGUYET

Hai Phong University

Email: conguyetdhhp1968@gmail.com

NGUYEN THI THUY HANG

Faculty of Applied Science, Ho Chi Minh City University of Technology, 268 Ly Thuong Kiet, District 10

Email: hangthuynguyen2001@yahoo.com

NGUYEN THI VAN OANH

University of Paris-Sud

Email: van-oanh.nguyen-thi@u-psud.fr

NGUYEN THUY TRANG

Faculty of physics, Hanoi University of Science

Email: trangnguyenphys@gmail.com

NGUYEN TRI LAN

Institute of Physics, VAST

Email: ntlan@iop.vast.ac.vn

NGUYEN TRUONG CO

Institute for Computational Science and Technology at Ho Chi Minh City

Email: truongcophysics@gmail.com

NGUYỄN VĂN DUY

University of Science and Technology, Hanoi

Email: duy11nv@yahoo.com

NGUYEN VAN HOA

Ho Chi Minh City University of Pedagogy

Email: nvhoa62@yahoo.com

NGUYEN VAN HONG

Hanoi University of Science and Technology

Email: hong.nguyenvan@hust.edu.vn

NGUYEN VAN HUNG

Physics Department, University of Science, VNU, Hanoi

Email: hungnv@vnu.edu.vn

NGUYEN VAN KHOE

University of Science, National University of Hochiminh City

Email: nvkhoe@hcmus.edu.vn

NGUYEN VAN THANH

Physics Department, University of Science, VNU, Hanoi

Email: thanhpc1987@gmail.com

PHUOC THE NGUYEN

Duy Tan University

Email: thenpk14@gmail.com

PAOLO CARLONI

German Research School for Simulation Sciences GmbH, Julich, Germany

Email: p.carloni@grs-sim.de

PETER FULDE

Max-Planck-Institut für Physik komplexer Systeme, Dresden, Germany

Email: fulde@apctp.org

PHAM DANG LAN

Institute for Computational Science and Technology

Email: l72mss@gmail.com

PHAM DINH QUOC HUY

Institute of Computational Science and Technology, Ho Chi Minh city

Email: phamdinhquochuy@gmail.com

PHAM HUONG THAO

Faculty of Physics, Hanoi university of Science.

Email: hthao82@gmail.com

PHAM TIEN LAM

University of Transport and Communication, Hanoi, Vietnam

Email: lampt.vn@gmail.com

PHAN VAN NHAM

Institute of Physics, Vietnamese academy of Science and Technology

Email: pvnham@iop.vast.ac.vn

PHUNG NGUYEN THAI HANG

Trường Đại học Tây Nguyên, Đại học Khoa học Tự nhiên Tp. Hồ Chí Minh

Email: thaihang72@gmail.com

PHÙNG VĂN ĐỒNG

Institute of Physics, VAST

Email: pvdong@iop.vast.ac.vn

PURU JENA

Virginia Commonwealth University

Email: pjena@vcu.edu

QUANG KHA QUACH

Department of Biological Sciences, Pusan National University, Busan 609-735

Email: quachkhaquang@pusan.ac.kr

THUY ANH CHU

Institute of Physics, VAST

Email: ctanh@iop.vast.ac.vn

TAE-SOO CHON

Department of Biological Sciences, Pusan National University, Busan (Pusan), 609-735, Republic of Korea

Email: tschon@pusan.ac.kr

THI HAI YEN VU

Laboratoire des Solides Irradiés, École Polytechnique

Email: thi-hai-yen.vu@polytechnique.edu

TRAN MINH HIEN

Ho Chi Minh City Institute of Physics

Email: htdung8@yahoo.com

TRẦN MINH TIẾN

Institute of Physics, VAST

Email: minhtien@iop.vast.ac.vn

Journal of Physics: Conference Series **537** (2014) 011001 doi:10.1088/1742-6596/537/1/011001

TRAN PHUOC DUY

Faculty of Applied Sciences, Hochiminh City University of Technology

Email: tranpduy87@yahoo.com

TRAN VAN NAM

Hanoi University of Science

Email: namtran.phys@gmail.com

TRINH XUAN HOANG

Institute of Physics, VAST

Email: hoang@iop.vast.ac.vn

TRUONG MINH DUC

Đại hoc Sư Phạm Huế

Email: tmduc2009@gmail.com

VAN-NAM DO

Advanced Institute for Science and Technology, Hanoi University of Science and Technology

Email: vannam.do@gmail.com

VIET-HUNG NGUYEN

Institut Nanosciences et Cryogénie/CEA

Email: nvhungphys@gmail.com

VO VAN HOANG

Department of Physics, Institute of Technology, Vietnam National University- HochiMinh City

Email: vvhoang2002@yahoo.com

VO VAN ON

Đại Học Thủ Dầu Một

Email: vovanon1963@gmail.com

VU BA DUNG

Đại học Mỏ - Địa chất Hà Nội

Email: vubazung305@gmail.com

VU NGOC TUOC

Hanoi University of Science and Technology

Email: tuoc.vungoc@hust.edu.vn

VU THI NGOC HUYEN

Institute of Physics, VAST

Email: vnhuyen@iop.vast.ac.vn

YUAN PING FENG

Department of Physics, National University of Singapore

Email: phyfyp@nus.edu.sg

Journal of Physics: Conference Series **537** (2014) 011002 doi:10.1088/1742-6596/537/1/011002

Peer review statement

All papers published in this volume of *Journal of Physics: Conference Series* have been peer reviewed through processes administered by the proceedings Editors. Reviews were conducted by expert referees to the professional and scientific standards expected of a proceedings journal published by IOP Publishing.

Journal of Physics: Conference Series **537** (2014) 012001 doi:10.1088/1742-6596/537/1/012001

Theory and Simulation of Magnetic Materials: Physics at Phase Frontiers

H T Diep[1], Virgile Bocchetti[1], Danh-Tai Hoang[2] and V T Ngo[3]

[1] Laboratoire de Physique Théorique et Modélisation, Université de Cergy-Pontoise, CNRS, UMR 8089, 2 Avenue Adolphe Chauvin, F-95302 Cergy-Pontoise Cedex, France

[2] Asia Pacific Center for Theoretical Physics, POSTECH, San 31, Hyoja-dong, Nam-gu, Pohang, Gyeongbuk 790-784, Korea

[3] Institute of Physics, Vietnam Academy of Science and Technology, 10 Daotan, Thule, Badinh, Hanoi, Vietnam

E-mail: diep@u-cergy.fr, virgile.bocchetti@u-cergy.fr, danh-tai.hoang@apctp.org, nvthanh@iop.vast.ac.vn

Abstract. The combination of theory and simulation is necessary in the investigation of properties of complex systems where each method alone cannot do the task properly. Theory needs simulation to test ideas and to check approximations. Simulation needs theory for modeling and for understanding results coming out from computers. In this review, we give recent examples to illustrate this necessary combination in a few domains of interest such as frustrated spin systems, surface magnetism, spin transport and melting. Frustrated spin systems have been intensively studied for more than 30 years. Surface effects in magnetic materials have been widely investigated also in the last three decades. These fields are closely related to each other and their spectacular development is due to numerous applications. We confine ourselves to theoretical developments and numerical simulations on these subjects with emphasis on spectacular effects occurring at frontiers of different phases.

1. Introduction

The physics at frontiers of different phases of the same system is a very exciting subject. In general, when there exist between the particles of a system several interactions, each of which favors a different symmetry, the system chooses the symmetry where its internal energy is minimum if the temperature $T = 0$ or where its free energy is minimum if $T \neq 0$. Let us take a simple example: we consider a chain of Ising spins interacting with each other via an interaction J_1 between nearest neighbors (NN) and an interaction J_2 between next NN (NNN). If J_1 is ferromagnetic ($J_1 > 0$) and $J_2 = 0$ then the ground state (GS) is ferromagnetic. Now, if J_2 is antiferromagnetic ($J_2 < 0$), we cannot arrange the spins in an order that fully satisfies at the same time J_1 and J_2: if $|J_2| \ll (\gg) J_1$ then the system chooses to satisfy J_1 (J_2) by taking the ferromagnetic (antiferromagnetic) order. There exists a critical value η_c of J_2/J_1 where the system changes from the ferromagnetic symmetry to the antiferromagnetic one. The critical value η_c is the frontier between the two phases. In general, when there are more than two interactions, the determination of the frontiers between different phases is more complicated. As we see, the competition between "incompatible" interactions creates frontiers. This is not limited to physics. Near a frontier, physical behaviors are different on the two sides: due to their

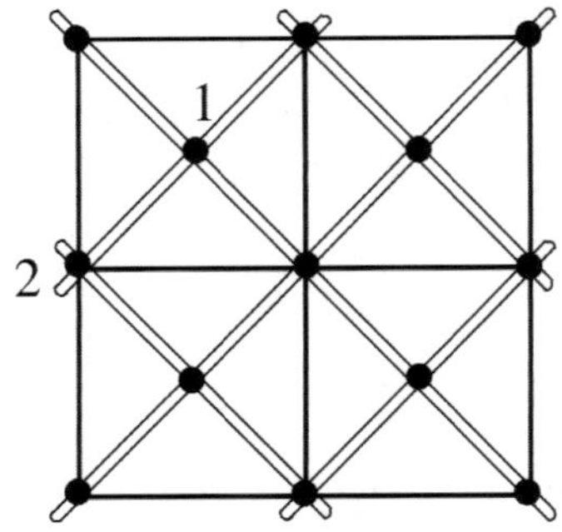

Figure 1. Centered square lattice. Interactions between NN and NNN, J_1 and J_2, are denoted by white and black bonds, respectively. The two sublattices are numbered 1 and 2.

different symmetries, they have different laws that govern fluctuations etc. When we introduce into one phase an external perturbation such as the temperature or an applied field, the system can choose the symmetry of the other phase. Such a phenomenon occurs very often in many systems: it is called "reentrance".

In this review, we give various examples of physical phenomena which occur around the phase frontiers. These examples are taken from our recent and current works.

The first subject concerns frustrated spin systems where competing interactions cause many spectacular effects (see reviews given in Ref. [1]). This is shown in section 2. We will describe there the reentrance, the disorder lines and the partial disorder which have been found in exactly solved models [2, 3, 4, 5]. We will show that simulations give complementary results that exact methods cannot reach. The second subject concerns theories and simulations of magnetic thin films. This is shown in section 3. One of the most important surface effects is the existence of localized spin-wave modes near a surface and interface. It has been shown a long time ago [6] that low-lying surface modes affect strongly macroscopic properties of magnetic systems giving rise for instance to a low surface magnetization or surface magnetically dead layer and to surface phase transitions at low temperatures. We discuss in particular surface effects in frustrated materials and in films with dipolar interaction [7, 8]. Recent results of spin resistivity are shown and discussed in relation with surface disordering [9, 10]. Finally, results on melting and surface lattice relaxation [11] are also shown. Concluding remarks are given in section 4.

2. Frustrated spin systems near phase frontiers

A system is said frustrated when all interaction bonds cannot be simultaneously satisfied in the GS. Well-known examples include the triangular lattice with NN antiferromagnetic interaction and the example given in the Introduction. For definiteness, let us take the case of the "centered square lattice" with Ising spins shown in Fig. 1, introduced by Vaks *et al.* [12] with NN and NNN interactions, J_1 and J_2, respectively. The exact expression for the free energy, some correlation functions, and the magnetization of one sublattice were given in the original work of Vaks *et al.*

The GS properties of this model are as follows : for $a = J_2/ \mid J_1 \mid > -1$, spins of sublattice 2 orders ferromagnetically and the spins of sublattice 1 are parallel (antiparallel) to the spins of sublattice 2 if $J_1 > 0$ (< 0); for $a < -1$, spins of sublattice 2 orders antiferromagnetically, leaving the centered spins free to flip. The phase diagram of this model is given by Vaks *et al.* [12]. Except at the "frontier" $a = -1$, there is always a finite critical temperature. When J_2 is antiferromagnetic (> 0) and J_2/J_1 is in a small region near 1, namely near the frontier of the two

Journal of Physics: Conference Series **537** (2014) 012001 doi:10.1088/1742-6596/537/1/012001

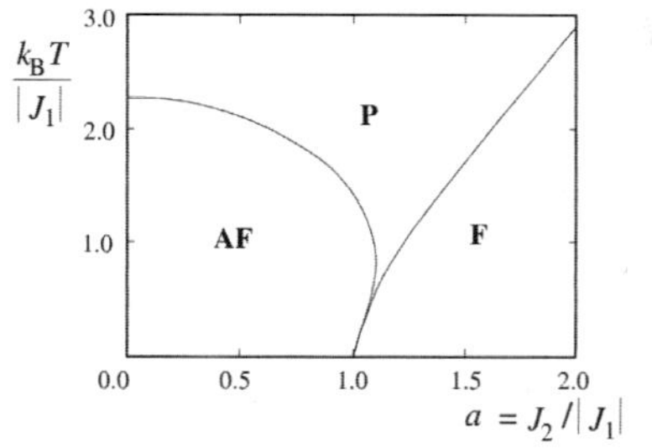

Figure 2. Phase diagram of centered square lattice near the "phase frontier" $a = -1$. [12]

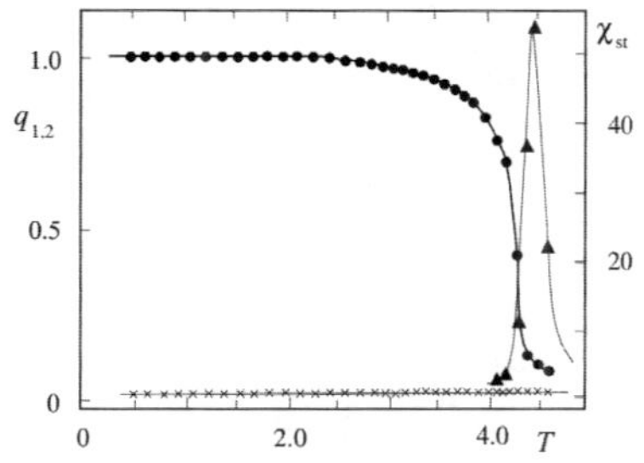

Figure 3. Temperature dependence of sublattice Edwards-Anderson order parameters, q_1 and q_2 (crosses and black circles, respectively) in the case $a = J_2/ \mid J_1 \mid = -2$, by Monte Carlo simulation. Susceptibility calculated by fluctuations of magnetization of sublattice 2 is also shown. The lattice used contains $N = 2 \times 60 \times 60$ spins with periodic boundary conditions. [13]

phases, the system is successively in the paramagnetic state, an ordered state, the **reentrant** paramagnetic state, and another ordered state, with decreasing temperature (see Fig. 2).

Though an exact critical line was obtained [12], the ordering in the antiferromagnetic (frustrated) region has not been exactly calculated. We have studied this aspect by means of Monte Carlo (MC) simulations [13] which show the coexistence between order and disorder. This behavior has been observed in three-dimensional (3D) Ising spin models [14, 15] and in an exactly soluble model (the Kagomé lattice) [2] as well as in frustrated 3D quantum spin systems [16, 17]. The results for the Edwards-Anderson sublattice order parameters q_i and the staggered susceptibility of sublattice 2 , as functions of T, are shown in Fig. 3 in the case $a = -2$.

As is seen, sublattice 2 is ordered up to the transition at T_c while sublattice 1 stays disordered at all T. This result shows a new example where order and disorder coexists in an equilibrium state. It noted that in the paramagnetic region, a Stephenson disorder line [18] has been found in Ref. [13]

$$\cosh(4J_1/k_BT_D) = \exp(-4J_2/k_BT_D) \tag{1}$$

The two-point correlation function at T_D between spins of sublattice 2 separated by a distance r is zero for odd distance r and decay like $r^{-1/2}[\tanh(J_2/k_BT_D)]^r$ for even r [18]. However,there is *no dimensional reduction* on the Stephenson line given above. Usually, one defines the disorder point as the temperature where there is an effective reduction of dimensionality so that physical

Journal of Physics: Conference Series **537** (2014) 012001 doi:10.1088/1742-6596/537/1/012001

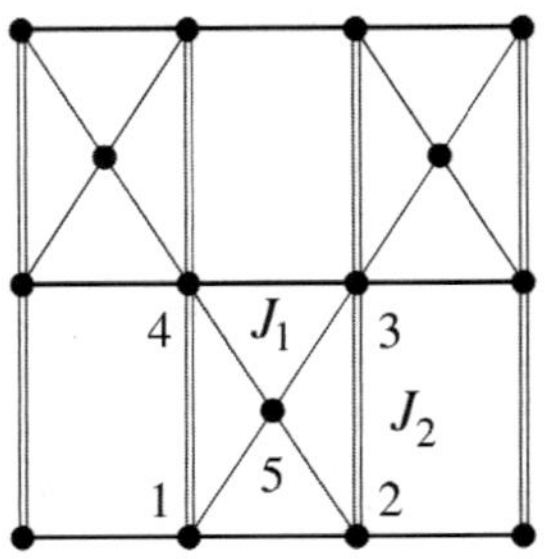

Figure 4. Kagomé lattice. Interactions between NN and between NNN, J_1 and J_2, are shown by single and double bonds, respectively.

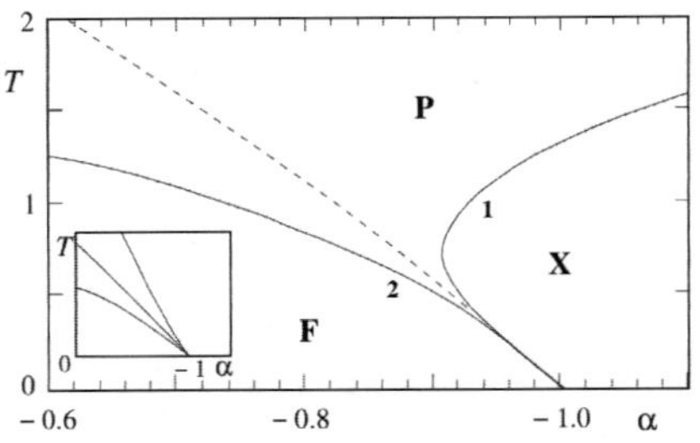

Figure 5. Phase diagram of the Kagomé lattice with NNN interaction in the region $J_1 > 0$ of the space ($\alpha = J_2/J_1, T$). T is measured in the unit of J_1/k_B. Solid lines are critical lines, dashed line is the disorder line. P, F and X stand for paramagnetic, ferromagnetic and partially disordered phases, respectively. The inset shows schematically enlarged region of the endpoint.

quantities become simplified spectacularly [19]. In general, these two types of disorder line are equivalent, as for example , in the case of the Kagomé lattice Ising model (see below). This is not the case here. The disorder line corresponding to dimensional reduction, was given for the general 8-vertex model by Ref. [20]. When this result is applied to the centered square lattice, one finds that the disorder variety is given by

$$\exp(4J_2/k_BT) = (1 - i\sinh(4J_1/k_BT))^{-1} \tag{2}$$

where $i^2 = -1$. This disorder line lies on the unphysical (complex) region of the parameter space of this system. Only the Stephenson disorder line Eq. (1) is the relevant one for the reentrance phenomenon. Disorder solutions have found interesting applications, as for example in the problem of cellular automata (for a review see Ref. [21]). Moreover, they also serve to built a new kind of series expansion for lattice spin systems [19].

Another example is the Kagomé lattice shown in Fig. 4. The phase diagram near the phase border is shown in Fig. 5 with the Stephenson disorder line. Other very rich phase diagrams are found in Refs. [2, 3, 4, 5] for honeycomb lattice, generalized Kagomé lattice and dilute centered square lattice. In some cases, up to five successive phase transitions and several disorder lines are found for a single set of interaction parameters.

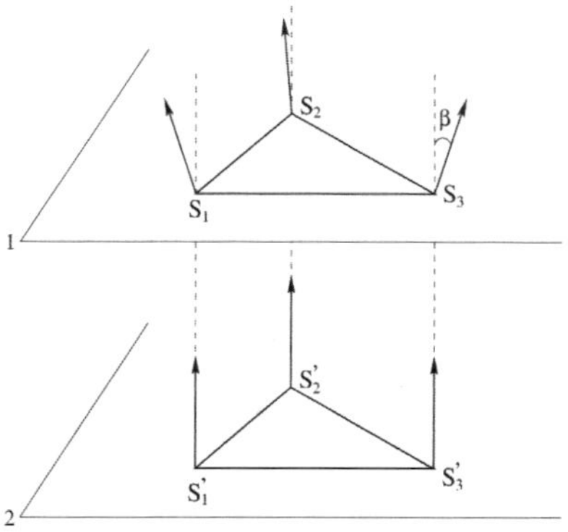

Figure 6. Non collinear surface spin configuration. Angles between spins on layer 1 are all equal (noted by α), while angles between vertical spins are β.

3. Magnetic thin films near the phase frontiers

The presence of a surface perturbs the bulk properties of a crystal. The perturbation becomes important when the ratio "number of surface atoms to number of bulk atoms" becomes significative. Among numerous surface effects, we will outline here some results on surface magnetization and surface phase transition near phase frontiers. The first example which illustrates very well the necessary combination between theory and simulation is the case of a FCC thin film with a (001) frustrated surface [7]. The Hamiltonian is given by

$$\mathcal{H} = -\sum_{\langle i,j \rangle} J_{i,j} \mathbf{S}_i \cdot \mathbf{S}_j - \sum_{<i,j>} I_{i,j} S_i^z S_j^z \quad (3)$$

where $\mathbf{S}_i$ is the Heisenberg spin at the lattice site i, $\sum_{\langle i,j \rangle}$ indicates the sum over the nearest neighbor spin pairs $\mathbf{S}_i$ and $\mathbf{S}_j$. The last term, which will be taken to be very small, is needed to ensure that there is a phase transition at a finite temperature for the film with a finite thickness when all exchange interactions $J_{i,j}$ are short-ranged. Otherwise, it is known that a strictly two-dimensional system with an isotropic non-Ising spin model (XY or Heisenberg model) does not have a long-range ordering at finite temperatures [22]. Between the surface spins we take $J_{i,j} = J_s$, and between all other spins $J_{i,j} = J = 1 > 0$ (ferromagnetic). The GS configuration depends on $\eta = J_s/J$. When η is smaller than a critical value, the GS becomes non linear. We show $\cos\alpha$ and $\cos\beta$ in Fig. 7 as functions of J_s where α and β are the angles between spins defined in the caption. The critical value J_s^c where the collinear configuration becomes non collinear is found between -0.18 and -0.19. This value can be calculated analytically (see [7]).

Using the theory of Green's function [6], we have calculated the layer magnetizations shown in Fig. 8 (see details in [7]). The phase diagram near the phase boundary J_s^c is shown in Fig. 9. We notice that the surface transition occurs only below J_s^c.

MC simulations give very similar results. Note however that at $T = 0$ quantum fluctuations causes a very strong "zero-point spin contraction" for surface spins: as a consequence, surface spins do not have the full length 1/2 as seen in Fig. 8, unlike the classical spins used in MC simulations.

The next example is a 2D layer with a dipolar interaction between the 3-state Potts model [8]. The competition is between the dipolar term of magnitude D, which favors the in-plane spin configuration, and the perpendicular anisotropy A which favors the perpendicular configuration. There is a critical value of D/A above (below) which the configuration is planar (perpendicular).

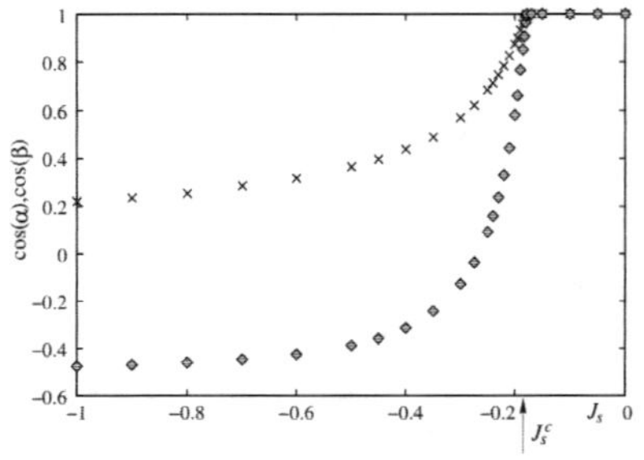

Figure 7. $\cos\alpha$ (diamonds) and $\cos\beta$ (crosses) as functions of J_s. Critical value of J_s^c is shown by the arrow.

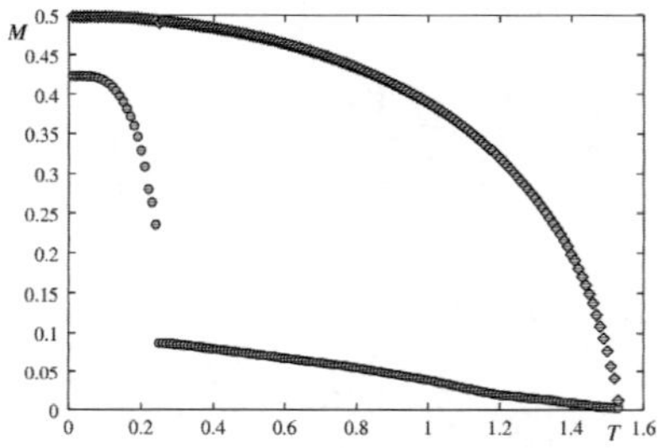

Figure 8. First two layer-magnetizations obtained by the Green's function technique vs. T for $J_s = -0.5$ with $I = -I_s = 0.1$. The surface-layer magnetization (lower curve) is much smaller than the second-layer one.

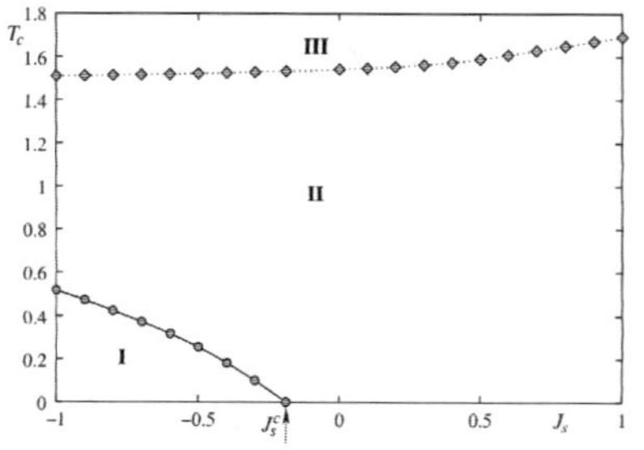

Figure 9. Phase diagram in the space (J_s, T) for the quantum Heisenberg model in a 4-layer film with $I = |I_s| = 0.1$. Phase I: all spins are ordered, phase II: for $J_s < J_s^c$ surface spins are disordered but bulk spins are ordered, for $J_s > J_s^c$ all spins are ordered, phase III: paramagnetic phase.

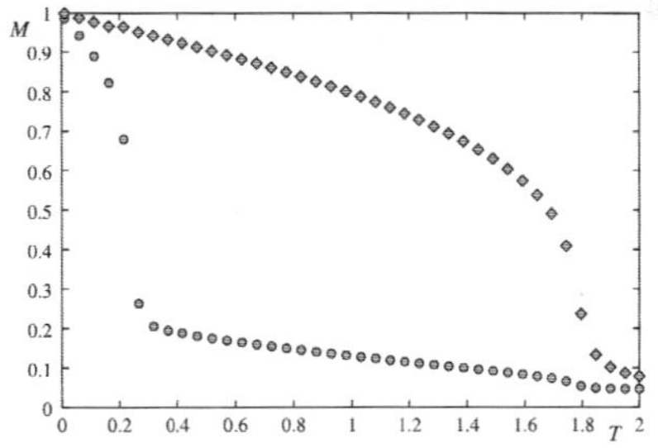

Figure 10. Monte Carlo results: magnetizations of layer 1 (circles) and layer 2 (diamonds) versus temperature T in unit of J/k_B for $J_s = -0.5$ with $I = -I_s = 0.1$.

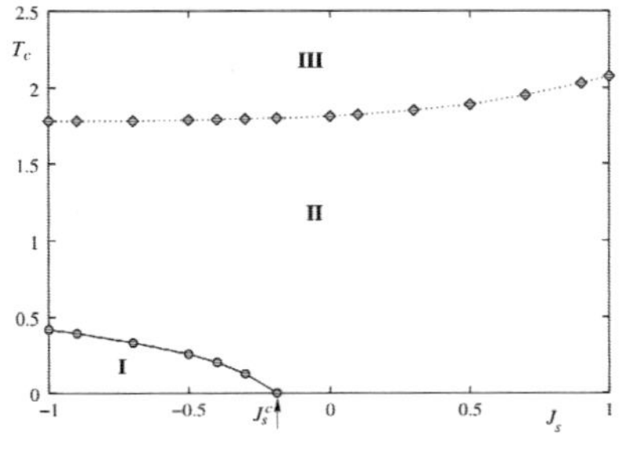

Figure 11. Phase diagram in the space (J_s, T) for the classical Heisenberg model in a 4-layer film with $I = |I_s| = 0.1$. Phases I to III have the same meanings as those in Fig. 9 .

Near this frontier, we found by MC simulations an interesting phenomenon which is called the re-orientation phase transition shown in Fig. 12: for example, if we follow the vertical line on the left figure, we see that at low T the system is in the planar configuration (phase II), it crosses the phase separation line with increasing T to go to the perpendicular phase (phase I) before going out to the paramagnetic phase (phase P) at a higher T. Such a re-orientation at a finite T is spectacular: we have shown that the transition between phases I and II is of first order [8]. This is not similar to the reentrant region shown in Figs. 2 and 5 where all transition lines are of second order and a very narrow paramagnetic phase separates the two ordered phases (AF and F, or F and X). It is interesting to note that to allow a transition between two "incompatible" symmetries (the one is not a subgroup of the other), there are two ways: i) a single first-order transition with a latent heat (as the case shown in Fig. 12), ii) two second-order transitions separated by a narrow region of a reentrant paramagnetic phase with often a disorder line separating two zones of different pre-ordering fluctuations (as the case shown in Fig. 5). Note that the re-orientation transition has also been observed for the Heisenberg spin model [23].

Let us give now an example where the surface disordering affects strongly the spin resistivity in a magnetic thin film. The spin resistivity has been studied both theoretically and experimentally for more than 50 years. The reader is referred to Refs. [9, 10, 24] for references. Theoretically, the coupling between an itinerant spin with lattice spins affects strongly the spin resistivity ρ. The spin-spin correlation has been shown to be the main mechanism which governs ρ. When there is a magnetic phase transition, the spin resistivity undergoes an anomaly. In magnetic thin films, when there is a surface phase transition at a temperature T_s different from that of the bulk one (T_c), we expect two peaks of ρ one at T_s and the other at T_c. We show here an example of a thin film, of FCC structure with Ising spins, composed of three sub-films: the

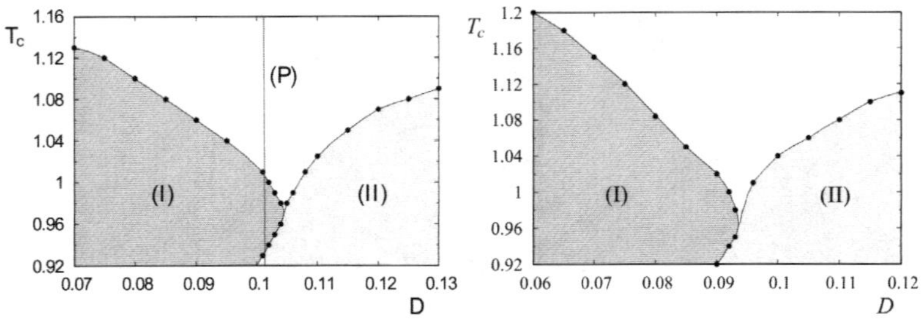

Figure 12. (Color online) Phase diagram in 2D: Transition temperature T_C versus D, with $A = 0.5$, $J = 1$ for two dipolar cutoff distances: $r_c = \sqrt{6}$ (left) and $r_c = 4$ (right). Phase (I) is the perpendicular spin configuration, phase (II) the in-plane spin configuration and phase (P) the paramagnetic phase.

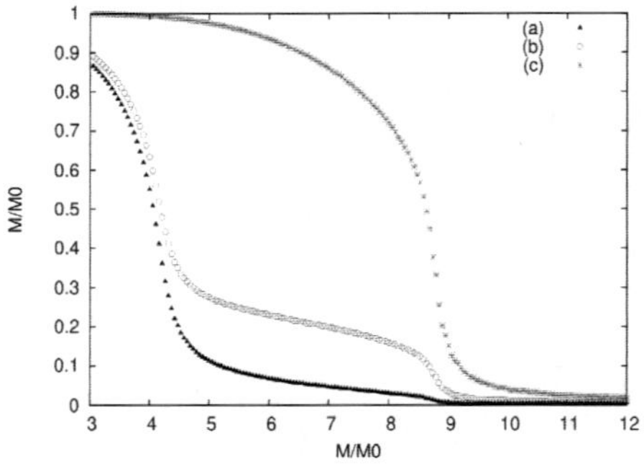

Figure 13. Magnetization versus T in the case where the system is made of three films (see text). Black triangles: magnetization of the surface films, stars: magnetization of the middle film, void circles: total magnetization.

middle film of 4 atomic layers between two surface films of 5 layers. The lattice sites are occupied by Ising spins interacting with each other via NN ferromagnetic interaction. Let us suppose the interaction between spins in the outside films be J_s and that in the middle film be J. The inter-film interaction is J. In order two enhance surface effects we suppose $J_s << J$. We show in Figs. 13 and 14 the layer magnetization and the spin resistivity for $J_s = 0.2J$. We observe that the surface films undergo a phase transition at $T_s \simeq 4$ far below the transition temperature of the middle film $T_c \simeq 8.8$. As stated above, a phase transition induces an anomaly in the spin resistivity: the two phase transitions observed in Fig. 13 give rise to two peaks of ρ shown in Fig. 14. The surface peak of ρ has been also seen in a frustrated film [10].

The last example is the surface relaxation and the surface melting of a semi-infinite Ag crystal with a (111) surface [11]. The border here, contrary to the previous examples, is a physical border (not a phase border). The competition exists also for this case because surface and bulk atoms have different environments and the multi-body interactions compete with the two-body terms in the potential from the Embedded-Atom-Method (EAM) [25] we have used. In order to see the surface melting, we compute the structure factor $S_{\vec{K}}$ as follows:

$$S_{\vec{K}} = \frac{1}{N_l} \left\langle \left| \sum_{j=1}^{N_l} e^{i\overrightarrow{K}.\overrightarrow{d_j}} \right| \right\rangle \tag{4}$$

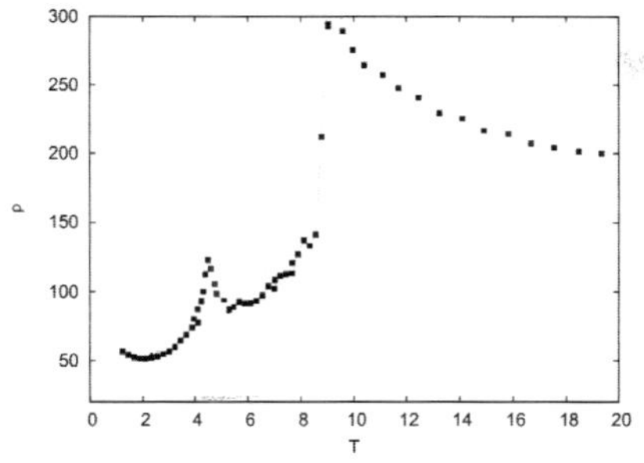

Figure 14. Resistivity ρ in arbitrary unit versus T of the system described in the previous figure's caption.

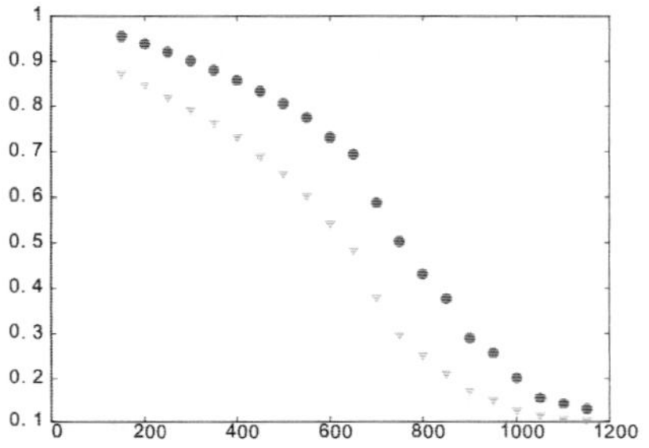

Figure 15. (Color online) Structure factor $S_{\vec{K}}$ (green triangles) and O_6 order parameter (red circles) of the first layer versus temperature for the EAM potential.

where $\vec{d_j}$ is the position vector of an atom in the layer, N_l the number of atoms in a layer and $\vec{K}$ the reciprocal lattice vector which has the following coordinate (in reduced units): $2\pi\left(1; -\sqrt{3}; 0\right)$. The angular brackets < ... > indicate thermal average taken over MC run time. The above "order parameter", which allows us to monitor the long-range surface order, is plotted for the surface layer in Fig. 15. As we can see, the long-range order is lost at $\simeq$ 700 K. Note that the bulk Ag melts at $\simeq$ 1235 K. In order to investigate in more details the surface melting, we have also computed the O_6 order parameter which describes the short-range hexatic orientational order of the surface :

$$O_6 = \frac{\left|\sum_{jk} W_{jk} e^{i6\Theta_{jk}}\right|}{\sum_{jk} W_{jk}} \tag{5}$$

with

$$W_{jk} = e^{-\frac{\left(z_j - z_k\right)^2}{2\delta^2}} \tag{6}$$

where the sum runs over the NN pairs and Θ_{jk} is the angle which the $j-k$ bond, when projected on the xy plane, forms with the x axis. The δ parameter is taken as one-half the average inter-layer spacing. The weighting function, W_{jk}, allows us to differentiate the "non coplanar" and the "coplanar" neighbors. With a coplanar neighbor, the weighting function takes a maximum value. We have to calculate the spatial average of O_6 taken over all atoms of the surface layer and then calculate its thermal average over MC run time. We plot the averaged O_6 parameter versus temperature in Fig. 15. The short-range order is also lost at 700 K. We have calculated the distance Δ between the topmost layer and the second layer. There is a contraction of this

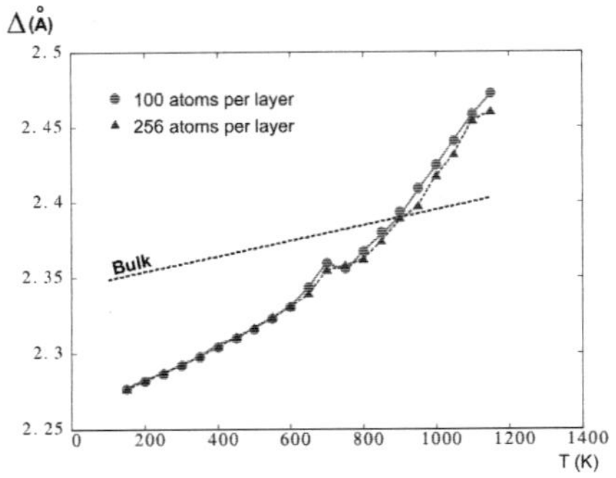

Figure 16. (Color online) Surface relaxation for two surface sizes 100 and 256 atoms for the EAM potential.

distance with respect to the distance between two layers in the bulk as seen in Fig. 16. Only at about 900 K, far after the surface melting, that surface atoms are "desorbed" from the crystal.

4. Concluding remarks

We have shown in this paper a number of phenomena occurring at frontiers between phases of different symmetries. The phase diagram near the phase frontiers is often very rich with reentrance, disorder lines and multiple phase transitions. One has seen that in most of the cases treated so far we need a combination of theory and simulation to better understand complicated physical effects resulting from competing forces around the frontiers. A frontier is determined as a compromise between these forces. As a consequence, frontiers do not have a high stability: we have seen that when a small external perturbation, such as the temperature, is introduced, the phase border moves in favor of one of the neighboring phases according to some criteria such as the entropy. Therefore, many interesting effects manifest themselves around frontiers. The physics near phase frontiers is far from being well understood in various situations.

Acknowledgments

H T Diep thanks the IOP-Hanoi for a financial support. V T Ngo is indebted to Vietnam National Foundation for Science and Technology Development (Nafosted) for the Grant No. 103.02-2011.55.

References

[1] Diep H T (ed) 2013 *Frustrated Spin Systems* 2nd edition (Singapore: World Scientific)
[2] Azaria P, Diep H T and Giacomini H 1987 *Phys. Rev. Lett.* **59** 1629
[3] Diep H T, Debauche M and Giacomini H 1991*Phys. Rev.* B **43** 8759
[4] Debauche M, Diep H T, Azaria P and Giacomini H 1991 *Phys. Rev.* B **44** 2369
[5] Debauche M and Diep H T 1992 *Phys. Rev.* B **46** 8214 ; Diep H T, Debauche M and Giacomini H 1992 *J. of Mag. and Mag. Mater.* **104** 184
[6] Diep-The-Hung, Levy J C S and Nagai O 1979 *Phys. Stat. Sol.* (b) **93** 351
[7] Ngo V T and Diep H T 2007 *Phys. Rev.* B **75** 035412
[8] Hoang D-T, Kasperski M, Puszkarski H, and Diep H T 2013 *J. Phys.: Cond. Matter* **25** 056006
[9] Akabli K and Diep H T 2008 *Phys. Rev.* B **77** 165433
[10] Magnin Y, Akabli K and Diep H T 2011 *Phys. Rev.* B **83** 144406
[11] Bocchetti V and Diep H T 2013 *Surf. Sci.* **614** 46
[12] Vaks V, Larkin A and Ovchinnikov Y 1966 *Sov. Phys. JEPT* **22** 820
[13] Azaria P, Diep H T and Giacomini H 1989 *Phys. Rev.* B **39** 740
[14] Blankschtein D, Ma M and Berker A N 1984 *Phys. Rev.* B **30** 1362
[15] Diep H T, Lallemand P and Nagai O 1985 *J. Phys.* C **18** 1067

Journal of Physics: Conference Series **537** (2014) 012001 doi:10.1088/1742-6596/537/1/012001

[16] Quartu R and Diep H T 1997 *Phys. Rev.* B **55** 2975
[17] Santamaria C and Diep H T 1997 *J. Appl. Phys.* **81** 5276
[18] Stephenson J 1970 *J. Math. Phys.* **11** 420 ; ibid 1970 *Can. J. Phys.* **48** 2118 ; ibid. 1970 *Phys. Rev.* B **1** 4405
[19] Maillard J M 1986 *Second Conference on Statistical Mechanics*, California Davies, unpublished.
[20] Giacomini H 1986 *J. Phys. A* **19** L335
[21] Rujan P 1987 *J. Stat. Phys.* **49** 139
[22] Mermin N D and Wagner H 1996, *Phys. Rev. Lett.* **17** 1133
[23] Santamaria C and Diep H T 2000 *J. Mag. Mag. Mater.* **212** 23
[24] Magnin Y and Diep H T 2012 *Phys. Rev.* B **85** 184413
[25] Zhou X W, Wadley H N G, Johnson R A, Larson D J, Tabat N, Cerezo A, Petford-Long A K, Smith G D W, Clifton P H, Martens R L and Kelly T F 2001 *Acta Materialia* **49** 4005

First-Principles Study on Crystal Phase Superlattice Nanowires Heterostructures

Vu Ngoc Tuoc[1], **Tran Doan Huan**[1,2], **Nguyen Viet Minh**[1], **Le Thi Hong Lien**[1]

[1] Institute of Engineering Physics, Hanoi University of Science and Technology, 1 Dai Co Viet Road, Hanoi, Vietnam

[2] Department of Physics, University of Basel, Klingelbergstrasse 82, CH-4056 Basel, Switzerland

E-mail: tuoc.vungoc@hust.edu.vn

Abstract. We perform a first-principles density functional theory study on structural and electronic properties of a sery of crystal-phase heterostructure atomic-scale superlattice (SL) nanowires (NW) from GaN material, i.e. GaN wurtzite(WZ) /zincblende (ZB) material interface. The effects of surface/interface relaxation and surface stress which are absent in atomistic models are carefully taken into account. Structural properties, energy bands and electronic properties for a class of hexagonal wires with various period of SL structure and diameter size are discussed. Pseudo hydrogen atoms, i.e. hydrogen with partial charges, are used to passivate the dangling surface bonds, which remove the localized in-gap surface states and suppress the surface reconstructions. With this passivation procedure the band structure show the type II for all wires. While the electrical aspects of these SL nanowires are explored through density functional theory, their subsequent band structures are used to determine the thermoelectric properties via the Boltzmann transport theory. Finally, the thermoelectric propertys dependence on temperature is unveiled.

1. Introduction

Nitride-based hetero-structure structured nanowires (NWs), either core/shell or superlattice (SL) heterostructure, i.e. of which parts are made of different nitrides material and/or different crystal phase, are promising for electronic and optoelectronic device applications, e.g., high-efficient, multicolor light emitting diodes [1, 2], solar cells [3], lasers [4]. Experimental works [1, 2, 3, 4, 5, 6, 7, 8, 9] and theoretical works [10, 11, 12, 13, 14, 15] indicate that various physical properties of these structures, especially the optical properties, can be tuned by several experimentally attainable parameters, including the size of the core/shell or SL period and diameter. This is one of the compelling advantages which fuel the broad interest on the structures. For a nano semiconductor structure, the surface reconstruction induced by the dangling bonds on the surface is essentially important and dramatically affects the structural and the electronic properties [16, 17, 18, 19, 20, 21, 22]. In case of a GaN SL crystal phase heterostructure nanowire, i.e. parts of wire with different structural phases along the nanowire direction, such as wurtzite (WZ) and zincblende (ZB) forms of the same compound, the WZ-ZB interface is another important factor. In addition, the quantum confinement in the SL NW is related to both the surface and the interface because the SL interface often form a type II band-alignment (staggered gap) with electrons on ZB and holes on WZ [5, 23], thus for the carriers which are already confined by the surface, residing aside the interface is favorable. The WZ-ZB phase lattice mismatch, 1% on the plane perpendicular to the WZ wire growth direction [0001] [23], also plays a role. All of these factors, simultaneously existing in the GaN crystal phase heterostructure SL NWs, result in combined effects on the properties of the structure. We present in this paper a first principles study on the role of the surface/interface reconstruction in GaN WZ-ZB SL nanowires. Further by a combination of first-principles simulations, based on the density functional theory, and Boltzmanns semiclassical theory, we have calculated some transport and thermoelectric properties of the this SL NW.

This paper is organized as follow. In Section 2, we describe the SL NWs and a reference model in which the surface effects are suppressed. Details on the numerical method used in this works are then introduced in Section 3. Section 4 is used to justify the reference model with bandstructure and Density-Of-States (DOS) analysis. In Section 5, we discuss the obtained results for bandstructure derived the transport and thermoelectric properties and the role of the surface reconstruction. The paper is closed by Section 6 with some conclusions.

2. GaN/AlN SL NWs and their passivated NW model

The most common crystal phase heterostructure consists of hexagonal wurtzite (WZ) and cubic zincblende (ZB) polytypes. Resulting heterostructures are typical of type-II, i.e., a ZB inclusion inside a WZ matrix forms a quantum well confining electrons, while a WZ inclusion in a ZB matrix forms a quantum well confining holes. It's noted that in the bulk materials, where inclusions of a different crystal polytypes inside a 3D matrix cannot be built and are typically surrounded by dislocations. In SL NW, crystal-phase heterostructures present several advantages with respect to compositional (i.e. from different materials) ones: (i) the abruptness of the heterointerface, coinciding with a *single basal plane*, (ii) chemical homogeneity, and (iii) the low lattice mismatch. Basal SFs can be regarded as a single cubic sequence of planes (ABC) in the hexagonal matrix ($ABAB...$) and are associated with a strong luminescence signal due to bound excitons at energy $3.41 - 3.42eV$. For GaN SL NW the presence of ZB inclusions of different thickness in WZ GaN has only been reported in [24] and an analysis of thermo-electric of such structures has not been carried out.

The building block of the GaN WZ layers in this work is a WZ hexagonal base layer, which is being composed of several hexagonal rings of atoms enclosing the NWs axis, taken to be parallel to the WZ [001] direction (ab stacking sequence) or as z axis. A WZ hexagonal base consist of 24 atoms with diameter of $10.1\AA$ (Fig1a). For ZB base layer a hexagonal shape of [111] cut-plane (abc stacking sequence) with suitable size is taken, so that a ZB hexagonal base consist of 25 atoms with diameter of $11.7\AA$ (Fig1b). These shapes of nanowires are chosen in such a way as to minimize the number of dangling bonds on outer edge-layer atoms, that for WZ domain the outer layer atoms have a minimum coordination of three, leaving at most one dangling bond on the edge atoms. For ZB domain there are three hexagonal edge atoms with coordination number of two leaving two dangling bonds for each atomic layer (see last image on Fig. 1). There three edge Ga atoms on one atomic layer following by three N edge atoms on the next atomic layer successively, thus making a dangling bond density on the ZB region is significantly higher than in the WZ region. This lead to more contribution of ZB domain on the surface state and surface reconstruction.

It is mentioned that for a NW, the surface reconstruction is an important factor that strongly affects physical properties of the NWs. To estimate the effects from the surface reconstruction, for each NW, we study a reference NW in which a number of pseudo hydrogen atoms are used to saturate the dangling bonds, thus suppressing the surface reconstruction and thus maintaining nearly bulk bond angle and bond length for the outer surface atom layer. These are artificial hydrogenic atom with partial charge. i.e. a charge of $3/4e$ or $5/4e$ with e is electron charge,

Journal of Physics: Conference Series **537** (2014) 012002 doi:10.1088/1742-6596/537/1/012002

depending on whether the dangling bond is on a nitrogen or gallium atom, respectively [22]. We call the reference NWs by *passivated* NWs while the original NWs are called *unpassivated* NWs. To passivate a WZ base layer one needs 12 suitable pseudo hydrogen atoms whereas for a ZB layer needs 15 atoms.

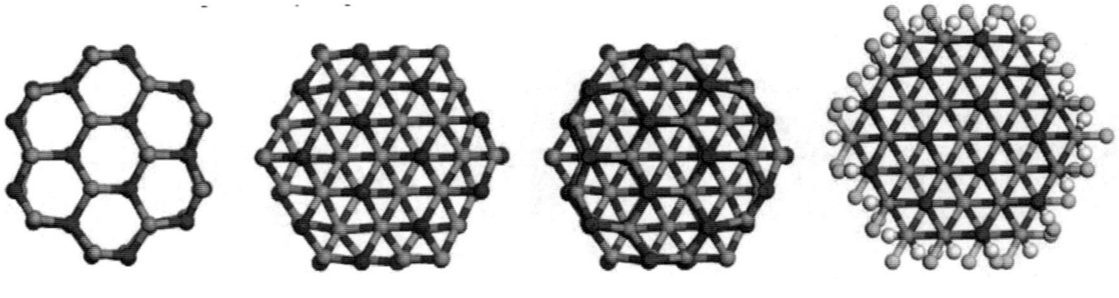

Figure 1. (Color online) From left to right: atomic scheme of relaxed GaN wurtzite (WZ) base layer; zinc-blende (ZB) base layer; real SL insertion WZ-ZB and a passivated SL NW with pseudo hydrogen. Beige sphere for Galium atom, Blue and Cyan spheres correspond to N atoms in the WZ and ZB domains, respectively whereas white is pseudo hydrogen with partial charge (3/4) and light blue is pseudo hydrogen with partial charge (5/4) for passivation the dangling bonds.

Table 1. Summary of structure's series of SL nanowires studied in this work, i.e., SL NW: A,B,C and their passivated SL NW: AH, BH, CH.

Structure	Number atoms	Relaxed Length (Å)	Ratio WZ:ZB layer
A	122	13.13	3:2
B	170	18.90	5:2
C	196	21.7	4:4
AH	188	13.59	—
BH	260	19.01	—
CH	304	21.73	—

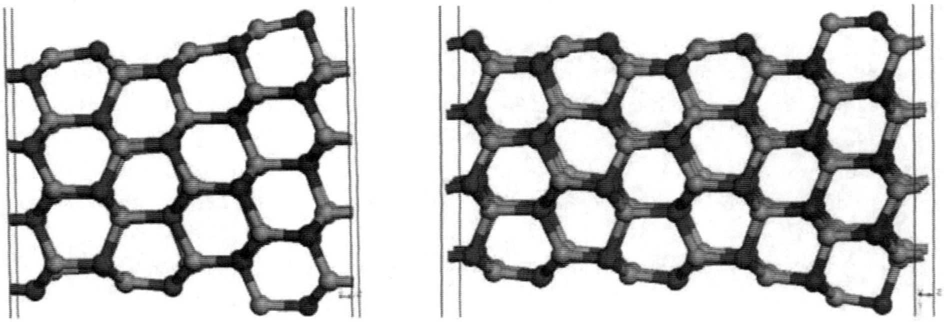

Figure 2. (Color online) Relaxed unpassivated SL NW A and B viewed along wire direction. Color scheme as in Fig.1.

Journal of Physics: Conference Series **537** (2014) 012002 doi:10.1088/1742-6596/537/1/012002

3. Computational details

We use in this work the version of density functional theory (DFT) implemented in SIESTA (Spanish Initiative for Electronic Simulations with Thousands of Atoms), the electronic structure code that implements linear-scaling methods for gapped systems [25, 26].The flexible localized numerical double-zeta-plus polarization basis set is used, allowing arbitrary angular momenta, multiple radial functions per angular momentum, and polarized and off-site orbitals [26]. Some figures in this work are prepared by XCrySDen [27]. Norm-conserving Troullier-Martins nonlocal pseudopotentials with partial core corrections are used for our calculations. The valence electron configurations taken for nitrogen, and gallium atoms are $3s2$ $3p1$ and $4s2$ $3d10$ $4p1$. Of these configurations, the gallium-$3d$ electron states are treated as valence states to describe the hybridization between the gallium-$3d$ and nitrogen-$2p$ orbitals. The pseudopotential radii employed in our calculations for s, p, d and f orbitals of nitrogen are all 1.14. Those for gallium are 2.1 au, 2.29 au, 2.1 au and 2.5 au, respectively. The pseudopotentials for the pseudo hydrogen atoms are generated from the pseudopotential of ordinary hydrogen atom using the *fractional* program implemented in Virtual Crystal Approximation, a tool provided by SIESTA 3.2, in which the pseudo hydrogen atoms are treated as the *synthesized* atoms. The Perdew-Burke-Ernzerhof functional [28] is used for exchange correlation energy of all the species.

The SL NWs structures are then optimized by minimizing the interatomic forces using selectively the conjugate gradient and the modified Broyden methods. The ground state energy and the optimized geometries of all NWs are carefully tested for convergence with respect to the size of the localized basis set and the dimensions of the supercell. The convergence is assumed when the forces on the atomic sites are less than $0.01eV/\AA$. The shift-energy parameter related to the cutoff radii of the localized atomic orbitals is chosen to be 150 meV for obtaining the optimized theoretical WZ lattice parameter for GaN ($a = 3.184\AA$ and $c = 5.186\AA$). These values are consistent with the experimental lattice parameters of real GaN ($a = 3.189\AA$ $c = 5.185\AA$) [23] For each NW, a system of two WZ and ZB domain in a box with the periodic boundary condition on the z direction. The structure is wrapped by at least $10\AA$ of vacuum thickness along the x and y directions to avoid the interaction of the system with its periodical images. The box size in the z direction, which is the lattice parameter c, is determined independently to minimize the total energy of the system.

Benchmarks of the passivated NW model, constructed with the pseudo hydrogen atoms, each of which is characterized by an artificial fractional charge, is used as the reference structure of which there is no surface reconstruction. We discuss some justifications for this model on detail in other work [22].

4. Electronic bandstructure and projected density-ofstate (PDOS) analysis

From the relaxed structures obtained in the previous Section, we have calculated the band structures and DOS of those SL NWs. To figure out the role and influence of the passivation procedure on Figure 3, the band structure of the unpassivated (Fig.3a) and passivated (Fig.3b) GaN-Wz nanowire are shown. Comparing to those of the passivated NW, the conduction band and the valence band of the unpassivated GaN-WZ are reconstructed with several new surface-induced states. For the conduction band, these states reside well just above the conduction band minimum (CBM) while for the valence band, the corresponding states are introduced right in the valence band maximum (VBM). In addition to the surface-induced states,there are also some small negative shifts on the other states of both the valence and the conduction bands. As a consequence of the surface-induced states, the band gap of the unpassivated GaN-WZ is dramatically reduced. i.e. from $2.51eV$ to $1.78eV$ (see Fig3c for *Pure WZ* column on the right), as also addressed elsewhere, e.g., in Ref. [22, 23]. The Fig.3c also show the gap dependence on the WZ-ZB composition ratio of all the studied sample. For the passivated NW (upper solid line in Fig.3c) it well restored the Vegard's rule for the bulk alloy composition, which corresponds to the case without surface effects. However for the unpassivated case, i.e. with surface effect the situation is totally changed as there is introduction of many surface states inside the gap make that small diameter NW structure become almost gap-less. Here we should note that although DFT calculations systematically underestimate the band gap of semiconductors, a discussion on the band gap of the NWs here is reasonable.

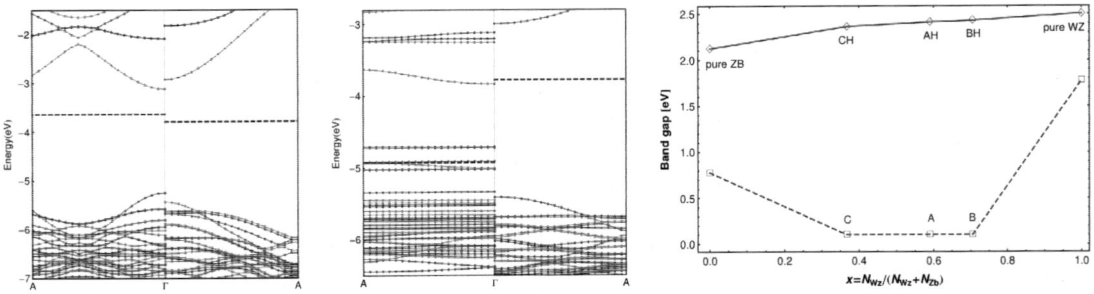

Figure 3. (Color online) Band structure of the unpassivated GaN-WZ nanowire (a) and passivated GaN-WZ NW (b). Dashed black lines present the Fermi levels of the NWs. Dotted-line rectangles on the figure indicate the surface-induced states.

The prominent effect can be noticed here is band gap's reduction due to the surface states as described above. The more surface states, i.e. depend on surface/volume as well as WZ/ZB ratio, the more band gap reductions. In Fig.4a the overlaying total DOS of unpassivated and passivated of a sample A was given, which shows that the sharp peak seen in the DOS of the NW is reduced from unpassivated to passived case and the surface states reside both the VBM and CBM and mid-gap states. For the nature of the surface-induced states, we examine further the projected density of states (PDOS), i.e., the electronic density of states projected on particular orbitals or atom sites of the NWs. In Fig. 4b, plots for the density of states of the passivated and unpassivated (upper window) and passivated (lower window) projected on particular domain (WZ and ZB) and orbitals are shown. It reveals that because of the surface reconstruction, both the valence and conduction band are changed dramatically by surface-induced states. For the unpassivated NW, the surface states marked by the arrows are dominated at both CBM and VBM regions. In particular, the surface state from ZB domain, i.e. $Ga - ZB$ and $N - ZB$ are dominated (see Fig. 4B upper window). For VBM edge states $N - ZB$ is more favor than $Ga - Zn$, where as for CBM edges the second come over. Within this the nitrogen $2p$ states dominate the VBM of N-ZB while contributions from gallium $4p$ states dominate the CBM with $Ga - ZB$ and nitrogen $2s$ of $N - ZB$ and $N - WZ$ $2p$ and $Ga - WZ$ $2p$ states are comparable in the CBM. For passivated NW, while the surface states are suppressed, the edges of VBM is dominated by the nitrogen $2p$ states of both $N - ZB$ and $N - WZ$ domains. For the CBM edges, Ga $4p$ of ZB domain almost dominated than N $2p$ ZB and N $2p$ and Ga $4p$ of WZ domain respectively (see Fig.4b lower window). Also it shows that without surface reconstruction, the gap gets wider and there are not a number of mid-gap states.

Figure 5 shows the spatial distribution of of two edges states at the VBM and CBM (the HOMO and LUMO) for unpassivated and passivated sample A. For passivated sample because the SL interface forms a heterostructure type II (stagged gap) of which both VBM and CBM of ZB-domain are smaller than of WZ-domain, edges state of the VBM are confined in the WZ-domain (left on second-row) where as edges state of CBM are confined in the ZB-domain (right image on second-row). The surface atoms make no contribution to these states since the surface reconstruction is suppressed by the present of pseudo hydrogen atoms, which have shown no contribution to VBM and CBM too.

The top row of Fig. 5 visualizes the dominant contribution of the surface-induced states to

Journal of Physics: Conference Series **537** (2014) 012002 doi:10.1088/1742-6596/537/1/012002

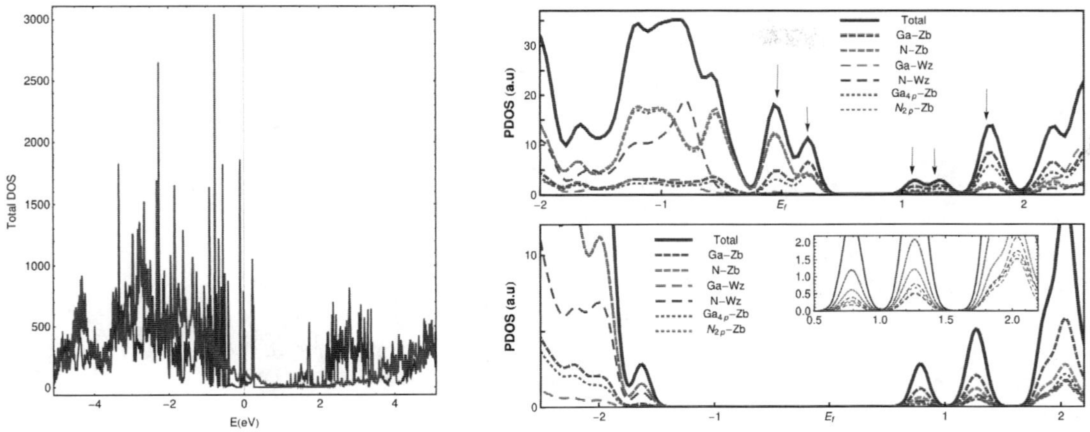

Figure 4. (Color online) (a) Total DOS of sample A (dark blue line) and its passivated - sample AH. (b) The Projected DOS of one sample A (unpassivated) and AH (passivated). Fermi levels is set to zero. The inset on lower window indicate the zooming of CBM edges.

the HOMO and LUMO as well as small contribution of atoms from core inside region. However it also can be seen that HOMO is more confined in the WZ region and LUMO in the ZB region as well. That result in a so-called *charge-separation*, i.e. when electronhole pairs will form by absorbed photon, this *charge separation* into electrons and holes at the interface's vicinity of the WZ and ZB stacking will occur, and electrons will be located in the ZB region and the holes will located in the WZ region (see Figure 5).

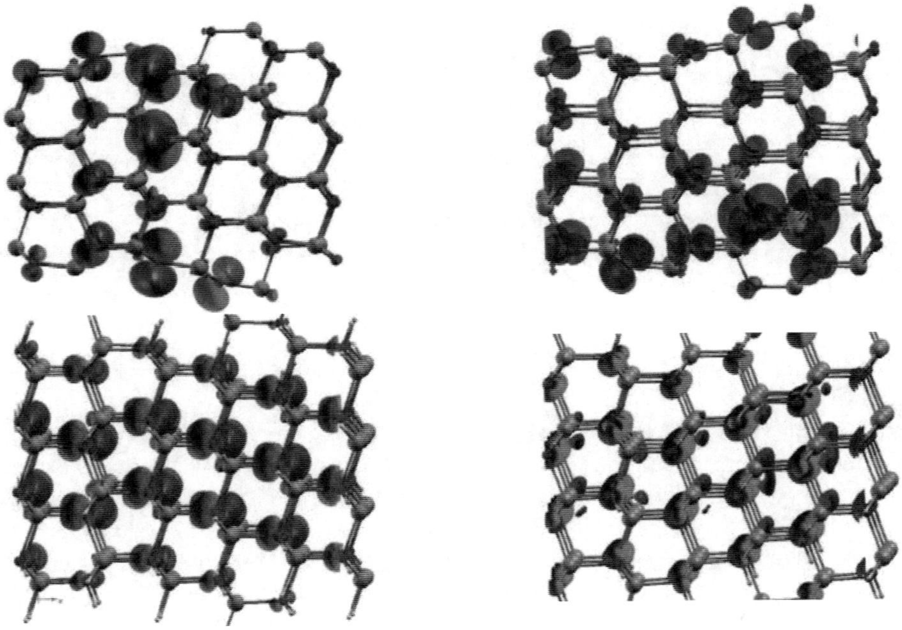

Figure 5. (Color online) Spatial distribution of HOMO and LUMO states of unpassivated sample A (top row) and passivated sample AH (bottom row). In this figure, galium and nitrogen atoms are shown in bedge and dark blue colors, respectively.

Journal of Physics: Conference Series **537** (2014) 012002 doi:10.1088/1742-6596/537/1/012002

5. Band-related transport properties

From this analytical representation of NW's band structure we calculate the derivatives necessary for the transport distributions. Based on the carefully calculated 1D bandstructure of SL NW along wire growth direction z, we have attempted to estimate the electrical conductivity σ and the Seebeck coefficient (also known as thermo-power) S and electronic contribution to the thermal conductivity κ_0 through the semiclassical Boltzmann's transport theory within the constant relaxation time approximation, as implemented in the BoltzTraP code [29]. This implementation relies on the smoothed Fourier expansion of the band's energies, provided by first-principles electronic structure calculation. i.e. by Siesta code at the above section. For the detail of the calculation we address to the similar scheme what was done in [30] for $SrTiO_3/SrRuO_3$ superlattices. As the group velocity is estimated from the first-principles calculations as a numerical derivative of band energies in a numerical mesh for the Brillouin zone sampling, there requires very fine grid. Since NW structures have only one infinite periodical dimensions of z, i.e. wire growth direction, so that only zz tensor component of transport properties are of interest. Fig.6a shows the temperature dependence of ZZ-component of tensor electrical conductivity per relaxation time σ_{zz}/τ of studied structure. All structures show similar behavour but the unpassivated (solid lines) and passivated (dashed lines) are different in 20 times in scaled showing the negative influence of surface confined states into conductivity and mobilities along wire. Similar tendency is also observed for ZZ-component of electronic thermal conductivity tensor κ_0 (Fig6.2) showing that the thermal conductivity is also affected by surface reconstruction and surface confined states. Finally the Seebeck coefficient tensor $S_z z$ of unpassivated wire have a sharp peak at low temperature before slowly monotonic increase as for the passivated case (see Fig.6.c).

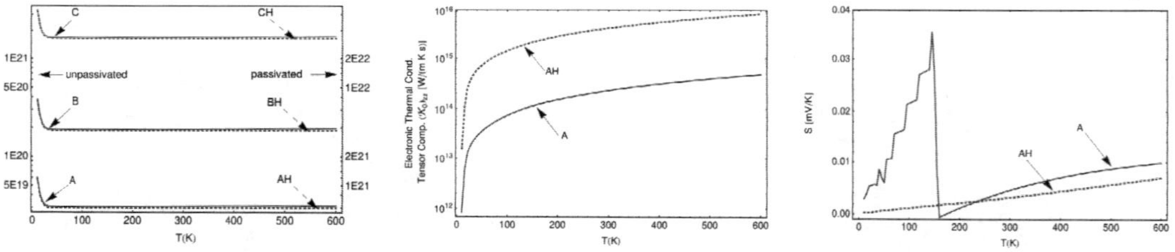

Figure 6. (Color online) (a) ZZ-component of electrical conductivity tensors σ_{zz}/τ as a function of relaxation time τ and temperature T (b) ZZ-component of electronic thermal conductivity tensor κ_0 vs T(c) ZZ-component of Seebeck coefficient tensor S_{zz} vs T.

6. Conclusions

We have presented in this work a study on the role of the surface reconstruction in several structures representing the GaN SL hetero- crystal-phase NWs with different size and composition. The effects of the surface reconstruction is estimated by the passivated NW model in which the dangling bonds on the NW's surface are saturated by fractionally charged pseudo hydrogen atoms. We have found that in this passivated NW model, as expected, effects from the surface reconstruction on the structural and electronic properties of the NWs are almost suppressed, leaving just a quantum confinement size effect. The structural and electronic properties of the NWs, i.e. Band and DOS, are strongly affected by the surface reconstruction. In particular, the band structure of the NWs is dramatically changed because of the surface reconstruction as there introduced a large number of *in-gap* states right in the vicinity of the VBM and the CBM. The WZ-Zb inteface form a type II quantum well and serves as charge separation for the excitonic pair and help to enhance the optical properties of SL NWs. The calculated transport properties of these SL NWS have also been strong affected by the present of the *in-gap* surface states.

Acknowledgments

Acknowledgments Part of this work (by V. N. T.) is supported by the Vietnamese NAFOSTED program 103.02-2011.20..

References

[1] T. Cohen-Karni, Q. Qing, X. Duan, P. Xie,1 C.M. Lieber, *Science* **329**, **830** (2010).
[2] F. Qian, S. Gradecak, Y. Li, C.-Y. Wen, and C. M. Lieber, *Nano Lett.* **5**, 2287 (2005).
[3] Y. Dong, B. Tian, T. J. Kempa and C.M. Lieber, *Nano Lett.* **9**, 2183 (2009).
[4] J. Arbio, S. Estrad, J. D Prades, A. Cirera, F. Furtmayr, C. Stark, A. Laufer ,M Stutzmann, M. Eickhoff, M. H Gass, A.L Bleloch, F. Peir and J. R Morante, *Nanotechnology* **20**, 145704, (2009).
[5] V Khranovskyy, A. M Glushenkov, Y Chen, A Khalid, H Zhang, L Hultman, B Monemar and R Yakimova, *Nanotechnology* **24**, 215202 (2013).
[6] L. Rigutti, G. Jacopin, L. Largeau, E. Galopin, A. De Luna Bugallo, F. H. Julien, J.-C. Harmand, F. Glas, and M. Tchernycheva, *Phys. Rev.* B **83**, 155320 (2011).
[7] V. Laneuville, F. Demangeot, R. P chou, P. Salles, A. Ponchet, G. Jacopin, L. Rigutti, A. de Luna Bugallo, M. Tchernycheva, F. H. Julien, K. March, L. F. Zagonel, and R. Songmuang, *Phys. Rev.* B **83** 115417 (2011).
[8] K. Hestroffer, R. Mata, D. Camacho, C. Leclere, G. Tourbot, Y. M. Niquet, A. Cros, C. Bougerol, H. Renevier and B. Daudin, *Nanotechnology* **21**, 415702 (2010).
[9] R. Calarco, M. Marso, T. Richter, A. I. Aykanat, R. Meijers, A. v.d. Hart, T. Stoica, and H. Luth, *Nano Letters* **5**, 981 (2005).
[10] B. Pal, K. Goto, M. Ikezawa, Y. Masumoto, P. Mohan, J.Motohisa, and T. Fukui, *Appl. Phys. Lett.* **93**, 073105 (2008).
[11] B. Goswami, S. Pal, and P. Sarkar, *J. Phys. Chem.* C **112**, 11630 (2008).
[12] J. Renard, R. Songmuang, G. Tourbot, C. Bougerol, B. Daudin, and B. Gayral, *Phys. Rev.* **80**, 121305 (2009).
[13] H. Lee and H. J. Choi, *Nano Lett.* **10**, 2207 (2010).
[14] R. N. Musin and X. Q. Wang, *Phys. Rev.* B **74**, 165308 (2006).
[15] L. Yang, R. N. Musin, X.Q. Wang, and M. Y. Chou, *Phys. Rev.* **77**, 195325 (2008).
[16] K. Rezouali, M. A. Belkhir, and J. B. Bai, *J. Phys. Chem.* **114**, 11352 (2010).
[17] Hui Pann, Yong-Wei Zhang *Nano Energy* **1**, 488493 (2012).
[18] B. Meyer and D. Marx, *Phys. Rev.* B **67**, 035403 (2003).
[19] N. Combe, P. Chassaing, and F. Demangeot, *Phys. Rev.* B **79**, 045408 (2009).
[20] V. N. Tuoc, *Comput. Mater. Sci.* **49**, S161 (2010).
[21] V. N. Tuoc and T. D. Huan, *Phys. Status Solidi B*, **249**, No. 3, 535543 (2012).
[22] V. N. Tuoc, T. D. Huan and L.T.H. Lien, *Phys. Status Solidi* B, **249**, No. 6, 12411249 (2012).
[23] L. Zhang, J. Luo, A. Zunger, N. Akopian, V. Zwiller, and J. Harmand, *Nano Lett.*, **10**, 4055-4060 (2010)
[24] G. Jacopin, L. Rigutti, L. Largeau, F. Fortuna, F. Furtmayr, F. H. Julien, M. Eickhoff, and M. Tchernycheva, *J. Applied. Phys.* **110**, 064313 (2011)
[25] P. Ordejon, E. Artacho, and J. M. Soler, *Phys. Rev.* B **53**, R10441 (1996).
[26] J. M. Soler, E. Artacho, J. D. Gale, A. Garcia, J. Junquera, P. Ordejon, and D. Sanchez-Portal, *J. Phys. Condens. Matter* **14**, 2745 (2002).
[27] A. Kokalj, *J. Mol. Graph. Model.* **17**, 176 (1999). Code available from http://www.xcrysden.org/.
[28] J. P. Perdew, K. Burke, and M. Ernzerhof, *Phys. Rev. Lett.* **77**, 3865 (1996).
[29] G. K. H. Madsen and D. J. Singh, *Comput. Phys. Commun.*, 175, **67** (2006).
[30] Garca-Fernndez, Pablo; Verissimo-Alves, Marcos; Bilc, Daniel I.; Ghosez, Philippe; Junquera, Javier *Phys. Rev.* B **86**, 085305 (2012).

Journal of Physics: Conference Series **537** (2014) 012003 doi:10.1088/1742-6596/537/1/012003

Photostimulated Radio Electrical Longitudinal Effect in a Parabolic Quantum Well

Bui Duc Hung[1], Nguyen Thi Thanh Nhan[1], Nguyen Quang Bau[1] and Nguyen Vu Nhan[2]

[1]Department of Physics, College of Natural Sciences, Hanoi National University, 334 Nguyen Trai, Thanh Xuan, Hanoi, Vietnam

[2]Department of physics, Academy of Defence force-Air force, Hanoi, Vietnam

E-mail: hungphysics.qn@gmail.com

Abstract. The longitudinal radioelectrical efffect in a parabolic quantum well (PQW) has been studied, based on the quantum kinetic equation for electrons under the action of a linearly polarized electromagnetic wave (EMW) and an intense laser field. Analytic expressions for the density of the current associated with the drag of charge carriers for the case of electron optical phonon scattering is calculated. The dependence of the current density on the intensity F and the frequency Ω of the laser radiation field, the frequency ω of the linearly polarized EMW field, the frequency ω_0 of the parabolic potential, the temperature T of the system are obtained. The analytic expressions are numerically evaluated and plotted for a specific quantum wells, GaAs/AlGaAs. All the results of PQW are compared with semiconductors builk and superlattice shows that the difference.

1. Introduction

In recent years, there has been considerable interest in the behavior of low-dimensional systems, in particular two-dimensional systems, such as semiconductor superlattices, doped superlattices and quantum wells. The confinement of electrons in low-dimensional systems considerably enhances the electron mobility and leads to unusual behaviors under external stimuli. Thus, under the action of EMW, the electrical conductivity, optical properties and kinetic effects are very different from semiconductors builk. Many papers have appeared dealing with problems related to the incidence of EMW [1] - [4]. The radioelectrical effect (the effect of drag of charge carriers by electromagnetic waves) is explained by the momentum transfer from photons to the electron, can be understood quasi-classically as being the result of the action of the Lorentz force on charge carriers moving in the ac electric and magnetic fields of the wave [5, 6]. The radioelectrical effect in isotropic semiconductors [7],in semiconductors [8, 9, 11, 13] have also been investigated and resulted by using the quantum kinetic equation for electrons system. In recent times, the radioelectrical effect in semiconductor superlattices has examined under the action of strong electric fields [16, 17] and of an elliptically polarized EMW [18]. However, the longitudinal radioelectrical efffect in a PQW still opens for studying. In this paper, we examine the system current carriers + scatterers, which is place in a constant electric field $\overrightarrow{E}_0$, a linearly polarized EMW field $\overrightarrow{E}(t) = \overrightarrow{E}(e^{-i\omega t} + e^{i\omega t})$ and in the presence of an intense laser field $\overrightarrow{F}(t) = \overrightarrow{F}\sin\Omega t$,in quantum wells with a parabolic potential. We consider the case in which the electron-optical phonon interaction is assumed to be dominant and electron gas to be nondegenerate. Numerical calculations are carried out with a specific GaAs/GaAsAl quantum wells. The comparison of the result of quantum wells to semiconductors builk and semiconductor superlattices shows that the difference.

2. Photostimulated Radio Electrical Longitudinal Effect in a Parabolic Quantum Well

2.1. The electron distribution function in PQW

In this paper, we apply the quantum kinetic equation method to study the longitudinal radioelectrical efffect in a PQW subjected to a dc electric field $\overrightarrow{E}_0 = (0, 0, E_{0z})$ and in a linearly polarized EMW field $\left(\overrightarrow{E}(t) = \overrightarrow{E}(e^{-i\omega t} + e^{i\omega t}), \vec{H}(t) = \left[\vec{n}, \overrightarrow{E}(t)\right]\right)$, ($\omega << \bar{\varepsilon}$, $\bar{\varepsilon}$ is an average carrier energy, in this paper, we select $\hbar = 1$) and in the presence of an intense laser field $\overrightarrow{F}(t) = \overrightarrow{F}\sin\Omega t$, $\Omega\tau << 1$ (τ is the characteristic relaxation time), the confinement potential being assumed to be $V(z) = m\omega_0 z^2/2$, then the motion of electrons is restricted in one dimension. The Hamiltonian of the electron-optical phonon system in the PQW in the second quantization representation can be written as [1-5] follows:

$$\begin{aligned} H &= \sum_{N,\vec{p}_\perp} \varepsilon_N(\vec{p}_\perp - \tfrac{e}{c}\overrightarrow{A}(t)) a^+_{N,\vec{p}_\perp} a_{N,\vec{p}_\perp} + \sum_{\vec{q}} \omega_{\vec{q}} b^+_{\vec{q}} b_{\vec{q}} \\ &+ \sum_{\substack{N,N' \\ \vec{p}_\perp,\vec{q}}} C_{\vec{q}} I_{N,N'}(q_z) a^+_{N',\vec{p}_\perp+\vec{q}_\perp} a_{N,\vec{p}_\perp} \left(b_{\vec{q}} + b^+_{-\vec{q}}\right) \end{aligned} \tag{1}$$

where N denotes the quantization of the energy spectrum in the z direction (N = 1,2,..), $|N, \vec{p}_\perp\rangle$ and $|N', \vec{p}_\perp + \vec{q}_\perp\rangle$ are electron states before and after scattering, $a^+_{N,\vec{p}_\perp}$ and $a_{N,\vec{p}_\perp}$ ($b^+_{\vec{q}}$ and $b_{\vec{q}}$)are the creation and annihilation operators of electron (phonon); $\omega_{\vec{q}}$ is the frequency of a phonon with the wave vector $\vec{q} = (\vec{q}_\perp, q_z)$; $\overrightarrow{A}(t)$ is the vector potential of laser field; $C_{\vec{q}}$is the electron-phonon interaction constant; $I_{N,N'}(q_z)$ is the electron form factor in the PQW. The electron energy takes the simple:

$$\varepsilon_N(\vec{p}_\perp) = \omega_p(N + \frac{1}{2}) + \frac{\vec{p}^2_\perp}{2m}(N = 0, 1, 2, ...) \tag{2}$$

with $\omega_p^2 = \omega_0^2 + \omega_H^2$ and $\omega_H = eH/mc$ are the confinement and the cyclotron frequencies.

2.2. Expressions for the drag of the charge carriers in the PQW.

The quantum kinetic equation for electrons in the constant scattering time (τ) approximation takes the form [5, 16, 17, 18]

$$\begin{aligned} &\tfrac{\partial f_{N,\vec{p}_\perp}(t)}{\partial t} - \left(e\overrightarrow{E}(t) + e\overrightarrow{E}_0 + \omega_H\left[\vec{p}_\perp, \vec{h}\right]\right) \tfrac{\partial f_{N,\vec{p}_\perp}(t)}{\partial \vec{p}_\perp} \\ &+ \tfrac{\vec{p}_\perp}{m}\tfrac{\partial f_{N,\vec{p}_\perp}(t)}{\partial \vec{r}} = -\tfrac{f_{N,\vec{p}_\perp}(t) - f_0(\varepsilon_{N,\vec{p}_\perp})}{\tau} \end{aligned} \tag{3}$$

where $\vec{h} = \frac{\overrightarrow{H}(t)}{H(t)}$ is the unit vector in the direction of magnetic field; $f_0(\varepsilon_{N,\vec{p}_\perp})$ is the equilibrium electron distribution function;$f_{N,\vec{p}_\perp}(t)$ is the nonequilibrium distribution function. In order to find $f_{N,\vec{p}_\perp}(t) = \left\langle a^+_{N,\vec{p}_\perp} a_{N,\vec{p}_\perp}\right\rangle_t$, we use the general quantum equation for the particle number operator or the electron distribution function

$$i\frac{\partial}{\partial t}f_{N,\vec{p}_\perp}(t)=\left\langle\left[a^+_{N,\vec{p}_\perp}a_{N,\vec{p}_\perp},H\right]\right\rangle_t \tag{4}$$

It is also assumed that the effect of the external field does not lead to a spatial inhomogeneity of the distribution function; i.e., $\frac{\partial f_{N,\vec{p}_\perp}(t)}{\partial\vec{r}}=0$. From Eqs. (3) and (4), using the Hamiltonian in Eq. (1), we obtain the quantum kinetic equation for electrons in the PQW

$$\begin{aligned}&-\left(e\vec{E}(t)+e\vec{E}_0+\omega_H\left[\vec{p}_\perp,\vec{h}(t)\right]\right)\frac{\partial f_{N,\vec{p}_\perp}(t)}{\partial\vec{p}_\perp}=-\frac{f_{N,\vec{p}_\perp}(t)-f_0(\varepsilon_{N,\vec{p}_\perp})}{\tau}+\\&+2\pi\sum_{N',\vec{q}}\left|C_{\vec{q}}I_{N,N'}(q_z)\right|^2\sum_{l=-\infty}^{+\infty}J_l^2(\vec{u},\vec{q}_\perp)\times\\&\left\{\left[f_{N',\vec{p}_\perp+\vec{q}_\perp}(t).(N_{\vec{q}}+1)-f_{N,\vec{p}_\perp}(t).N_{\vec{q}}\right].\delta\left(\varepsilon_{N',\vec{p}_\perp+\vec{q}_\perp}-\varepsilon_{N,\vec{p}_\perp}-\omega_{\vec{q}}-l\Omega\right)+\right.\\&\left.+\left[f_{N',\vec{p}_\perp-\vec{q}_\perp}(t).N_{\vec{q}}-f_{N,\vec{p}_\perp}(t).(N_{\vec{q}}+1)\right].\delta\left(\varepsilon_{N',\vec{p}_\perp-\vec{q}_\perp}-\varepsilon_{N,\vec{p}_\perp}+\omega_{\vec{q}}-l\Omega\right)\right\}\end{aligned} \tag{5}$$

in which $J_l(x)$ is the Bessel function of argument x;$\vec{u}=\frac{e\vec{F}}{m\Omega^2}$ is the amplitude of electron vibration in an EMW;$N_{\vec{q}}$ is the time-independent component of distribution function of phonons. For simplicity, we limit the problem to the case of l =0, ±1.We multiply both sides of Eq. (5) by $(-e/m)\vec{p}_\perp.\delta(\varepsilon-\varepsilon_{N,\vec{p}_\perp})$ and carry out the summation over N and $\vec{p}_\perp$. We obtain

$$\frac{\vec{R}_0(\varepsilon)}{\tau}=\vec{Q}_0(\varepsilon)+\vec{S}_0(\varepsilon)+\omega_H\left[\vec{R}(\varepsilon)+\vec{R}^*(\varepsilon),\vec{h}\right] \tag{6}$$

where

$$\vec{Q}_0(\varepsilon)=\frac{e}{m}\sum_{N,\vec{p}_\perp}\vec{p}_\perp\left(e\vec{E}_0,\frac{\partial f_0(\varepsilon_{N,\vec{p}_\perp})}{\partial\vec{p}_\perp}\right)\delta(\varepsilon-\varepsilon_{N,\vec{p}_\perp}) \tag{7}$$

and

$$\begin{aligned}\vec{S}_0(\varepsilon)=&-\frac{2\pi e}{m}\sum_{\substack{N,N'\\\vec{p}_\perp,\vec{q}}}\left|C_{\vec{q}}I_{N,N'}(q_z)\right|^2.N_{\vec{q}}\vec{q}_\perp.f_{10}(\vec{p}_\perp).\left\{\left[\left(1-\frac{(\vec{u}\vec{q}_\perp)^2}{2}\right)\times\right.\right.\\&\times\delta\left(\varepsilon_{N',\vec{p}_\perp+\vec{q}_\perp}-\varepsilon_{N,\vec{p}_\perp}-\omega_{\vec{q}}\right)+\frac{(\vec{u}\vec{q}_\perp)^2}{4}\delta\left(\varepsilon_{N',\vec{p}_\perp+\vec{q}_\perp}-\varepsilon_{N,\vec{p}_\perp}-\omega_{\vec{q}}+\Omega\right)+\\&\left.+\frac{(\vec{u}\vec{q}_\perp)^2}{4}\delta\left(\varepsilon_{N',\vec{p}_\perp+\vec{q}_\perp}-\varepsilon_{N,\vec{p}_\perp}-\omega_{\vec{q}}+\Omega\right)\right]-\left[\left(1-\frac{(\vec{u}\vec{q}_\perp)^2}{2}\right)\times\right.\\&\times\delta\left(\varepsilon_{N',\vec{p}_\perp-\vec{q}_\perp}-\varepsilon_{N,\vec{p}_\perp}+\omega_{\vec{q}}\right)+\frac{(\vec{u}\vec{q}_\perp)^2}{4}\delta\left(\varepsilon_{N',\vec{p}_\perp-\vec{q}_\perp}-\varepsilon_{N,\vec{p}_\perp}+\omega_{\vec{q}}+\Omega\right)+\\&\left.\left.+\frac{(\vec{u}\vec{q}_\perp)^2}{4}\delta\left(\varepsilon_{N',\vec{p}_\perp-\vec{q}_\perp}-\varepsilon_{N,\vec{p}_\perp}+\omega_{\vec{q}}-\Omega\right)\right]\right\}\delta\left(\varepsilon-\varepsilon_{N,\vec{p}_\perp}\right)\end{aligned} \tag{8}$$

and $\vec{R}_0(\varepsilon)$, $\vec{R}(\varepsilon)$ have meaning of the partial current densities with energy ε. These quantities are related to the total current density in the form

$$\vec{j}_{tot}=\vec{j}_0+\vec{j}(t)=\int_0^\infty\left\{\vec{R}_0(\varepsilon)+\left[\vec{R}(\varepsilon).e^{-i\omega t}+\vec{R}^*(\varepsilon).e^{i\omega t}\right]\right\}.d\varepsilon \tag{9}$$

We also consider the electron-optical phonon interaction. We consider the electron gas to be nondegenerate, thus, the electron distribution function is given by the Boltzmann distribution. In this case, $\omega_{\vec{q}} \ll \omega_{LO}$ is the frequency of the optical phonon in the equilibrium state, and the electron-phonon interaction constant is

$$|C_{\vec{q}}|^2 = \frac{2\pi e^2 \omega_{LO}}{\epsilon_0 q^2}\left(\frac{1}{\chi_\infty} - \frac{1}{\chi_0}\right) \tag{10}$$

with ϵ_0 is the permittivity of free space; χ_∞ and χ_0 are the static and the high-frequency dielectric constants, respectively. After some calculation Eq. (9) and taking the statistical average over time, we find the expression for the current density j_z along the 0z axis which expresses for the longitudinal radioelectrical efffect in a PQW

$$\begin{aligned}\langle j_z\rangle = &\left\{a + \tfrac{1}{4}\sqrt{\tfrac{m}{2\pi\beta}\tfrac{e^2F^2}{\Omega^4}}\cdot b\tau\right\}\tau.E_{0z}+ \\ &+\left\{a + \tfrac{1}{4}\sqrt{\tfrac{m}{2\pi\beta}\tfrac{e^2F^2}{\Omega^4}}\cdot b\tau\tfrac{\left[1-\omega^2\tau^2\right]}{1+\omega^2\tau^2}\right\}\tfrac{\omega_H\tau^2}{1+\omega^2\tau^2}E_x\end{aligned} \tag{11}$$

where $\beta = \frac{1}{k_BT}$ and

$$A = \frac{2\pi e^2\omega_{LO}}{m^2\epsilon_0 L_z}\left(\frac{1}{\chi_\infty} - \frac{1}{\chi_0}\right) \tag{12}$$

$$a = \frac{n_0^* e^2}{\pi\beta}\sum_N \exp\left\{-\beta\omega_0(N+\frac{1}{2})\right\} \tag{13}$$

$$b = A\frac{n_0^* e^2}{\pi\beta\omega_{LO}}\sum_{N,N'}(b_1+b_2+b_3+b_4+b_5+b_6) \tag{14}$$

$$b_1 = (2+\delta_{N,N'})B_1\exp\left\{-\beta\omega_0(N+\frac{1}{2})\right\}.\exp\left\{\frac{\beta B_1}{2}\right\}K_1\left\{\frac{\beta B_1}{2}\right\} \tag{15}$$

$$b_2 = (2+\delta_{N,N'})\frac{B_2}{2}\exp\left\{-\beta\omega_0(N+\frac{1}{2})\right\}.\exp\left\{\frac{\beta B_2}{2}\right\}K_1\left\{\frac{\beta B_2}{2}\right\} \tag{16}$$

$$\begin{aligned}&b_3 = b_2(B_2\to B_3); b_4 = -b_1(B_1\to -B_4);\\ &b_5 = -b_2(B_2\to -B_5); b_6 = -b_2(B_2\to -B_6);\end{aligned} \tag{17}$$

$$\begin{aligned}&b_3 = b_2(B_2\to B_3); b_4 = -b_1(B_1\to -B_4);\\ &b_5 = -b_2(B_2\to -B_5); b_6 = -b_2(B_2\to -B_6);\end{aligned} \tag{18}$$

Equation (11) shows the dependence of the current density on the intensity F and the frequency Ω of the laser radiation field, the frequency ω of the linearly polarized EMW field, the frequency ω_0 of the parabolic potential, the temperature T of the system. Now, we introduce the designations $j_0 = \mathrm{n_0 e}\varepsilon_F\mathrm{L_z}/\hbar$ and $I = (c/4\pi)E_x^2$ is the EMW intensity. Then, we will give a deeper insight into this dependence by carrying out a numerical assessment.

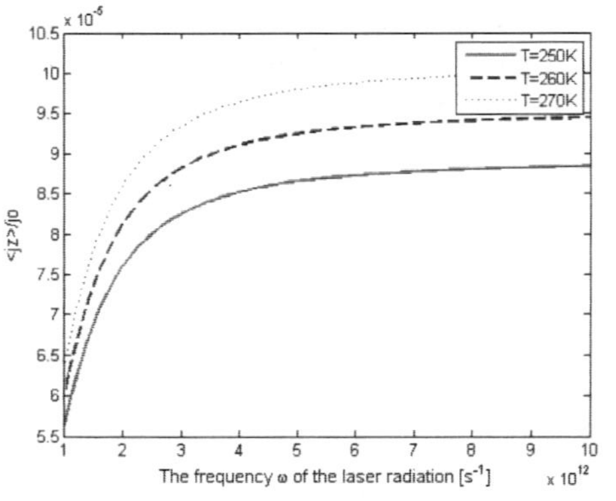

Figure 1. The dependence of $\langle j_z \rangle / j_0$ on the frequency ω of the EMW at $\Omega = 5.10^{14}(s^{-1})$, $F = 10^6(V/m)$, $\omega_0 = 0.5\omega_{LO}$ and $E_{0z} = -10^3(V/m)$

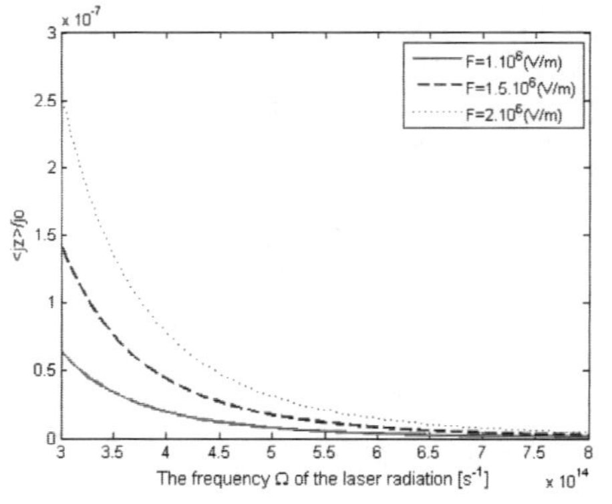

Figure 2. The dependence of $\langle j_z \rangle / j_0$ on the frequency Ω of the laser radiation field at $\omega = 10^{13}(s^{-1})$, $\omega_0 = 0.5\omega_{LO}$, T = 270K and $E_{0z} = -10^3(V/m)$

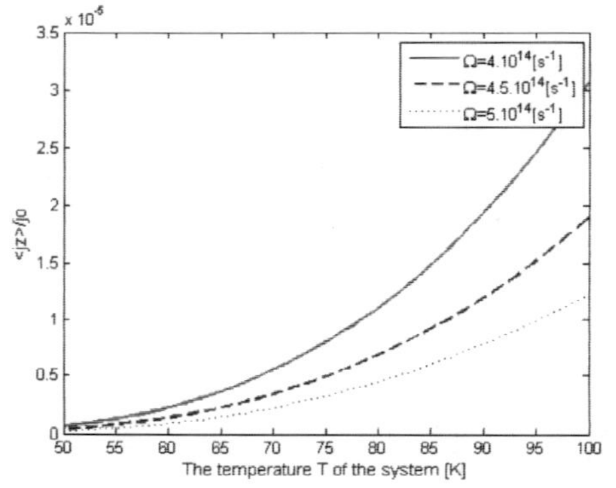

Figure 3. The dependence of $\langle j_z \rangle / j_0$ on T at $\omega = 10^{13}(s^{-1})$, $\omega_0 = 0.5\omega_{LO}$, $F = 2.10^6(V/m)$, and $E_{0z} = -10^3(V/m)$

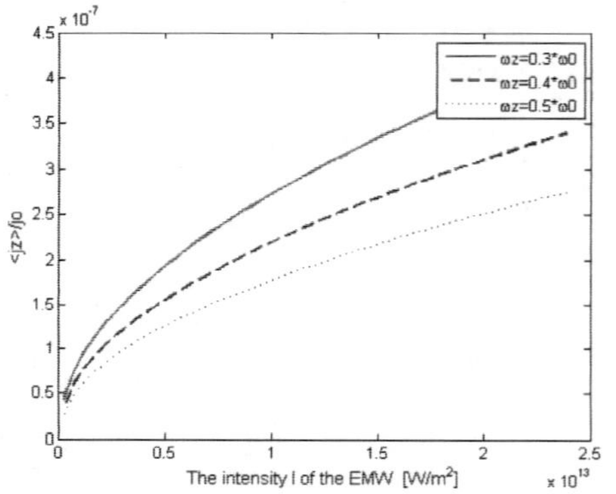

Figure 4. The dependence of $\langle j_z \rangle / j_0$ on I at $\Omega = 5.10^{14}(s^{-1})$, $F = 5.10^5(V/m)$, T = 260K and $E_{0z} = -10^5(V/m)$

2.3. Numerical results and discussion

In this section, we will evaluate, plot and discuss the drag of the charge carriers for the case of a specific GaAs/GaAsAl quantum wells. The parameters used in the calculations are as follows [5, 17, 18, 19]:$\epsilon_0 = 8,86.10^{-12}$; $\chi_\infty = 10{,}48$; $\chi_0 = 12{,}90$; $\hbar\omega_{LO} = 36{,}8$meV; m = 0,0665$m_0$ (m_0 is the mass of free electron); $e = 1,60219.10^{-19}C$; $\varepsilon_F = 50$meV; and we also choose $\tau(\varepsilon_F) \sim 10^{-11}s^{-1}$; $\tau(\Omega) \sim 10^{-10}s^{-1}$;

In Fig. 1 shows the dependence of the current density on the frequency ω of the EMW at different values of the temperature T of the system. From this figure, we can see that the current density increases strongly with increasing EMW frequency for the area of values $\omega < 5.10^{13}(s^{-1})$ and reaches saturation as the frequency ω continues to increase. Besides, the value of the current density raises remarkably when the temperature T increases.

The dependences of the current density on the intensity F and the frequency Ω of the laser radiation field are shown in Fig. 2. We can see that the values of the current density reduce nonlinearly when the frequency Ω increases. And the more amplitude F increases, the more the values of the current density raise up.

Fig. 3 shows that when the temperature T of the system rises up, the current density along the Oz axis goes up too. This figure confirms once again that the current density strongly depends on the frequency Ω of laser radiation.
Specially, Fig.4 shows the dependence of the current density on the intensity I of the EMW at different values of the confinement frequency ω_0 of the parabolic potential. This figure shows that the current density increases strongly with increasing the intensity I of the EMW. Moreover, the confinement of electrons in PQW also influences strongly on the current density along the Oz axis. The value of the current density reduces when the confinement frequency ω_0 increases.

3. Conclusion

In this paper, we have investigated the longitudinal radioelectrical efffect in a PQW subjected to a dc electric field, a linearly polarized EMW field and in the presence of an intense laser field. We obtain the expressions for the drag of the charge carriers in case the electron gas is nondegenerate. The dependences of the current density on the intensity F and the frequency Ω of the laser radiation field, the frequency ω of the linearly polarized EMW field, the frequency ω_0 of the parabolic potential, the temperature T of the system are obtained. The analytical results are numerically evaluated and plotted for a specific quantum wells, GaAs/AlGaAs, to confirm clearly once again that the current density strongly depends on the above elements.When $\omega_0 \to 0$, from (13, 14) we see a, b and $\langle j_z \rangle$ don't depend on ω_0, so the result will give back this effect in semiconductor [13]. The expression (11) is quite similar to the expressions in a superlattice in case $l = 0, \pm 1$ [18, 19]

Acknowledgment

This research is completed with financial support from the Program of Basic Research in National Foundation for Science and Technology Development (NAFOSTED, project No 103.01-2011.18).

References

[1] G. M. Shmelev, L. A. Chaikovskii and N. Q. Bau, *Sov. Phys. Semicond.* **12**, (1978) 1932.
[2] N. Q. Bau, D. M. Hung and L. T. Hung, *PIER Letters* **15**, (2010) 175.
[3] N. Q. Bau and D. M. Hung, *PIER B* **25**, (2010) 39.
[4] N. Q. Bau, D. M. Hung and N. B. Ngoc, *J. Korean Phys. Soc* **54**, (2009) 765.
[5] N. Q. Bau and B. D. Hoi, *J. Korean Phys. Soc.* **60**, (2012) 765.
[6] F. G. Bass, A. A. Bulgakov, and A. P. Tetervov, *High-Frequency Properties of Semiconductors with Superlattices* (1989) Nauka, Moscow.
[7] K. Seeger,*Semiconductor Physics* (1973)Springer-Verlag, Wien/New York.
[8] G. M. Shmelev, N. Q. Bau and N. H. Shon, *Izv. Vyssh. Uch. Zaved. Fiz.* **7** (1981) 105.
[9] V. L. Malevich and E. M. Epshtein *Izv. Vyssh. Uch. Zaved. Fiz.* **2** (1976) 121.
[10] V. L. Malevich *Izv. Vyssh. Uch. Zaved. RadioFizika.* **20** (1977) 151.
[11] N. H. Shon, G. M. Shmelev and E. M. Epshtein *Izv. Vyssh. Uch. Zaved. Fiz.* **5** (1984) 19.
[12] D. E. Milovzorov, *Technical Physics Letters*, **22** (1996) 896.
[13] G. M. Shmelev, G. I. Tsurkan and E. M. Epshtein, *Physica Status Solidi* B, **109** (1982) 53
[14] Cheng Wenqin, Huang Yi, Zhou Junming, Feng Wei, Xu Geng; *CPL*, **7** (1990) 284.
[15] G. M. Shmelev, N. H. Shon, G. I. Tsurkan, *Izv. Vyssh. Uch. Zaved. Fiz.* **2** (1985) 84.
[16] D. V. Zavyalov, S. V. Kryuchkov, and E. S. Sivashova, *Tech. Phys. Lett* **32** (2006) 143.
[17] D. V. Zavyalov, S. V. Kryuchkov, and E. I. Kukhar, *Semiconductors* **41** (2007) 704.
[18] S. V. Kryuchkov, E. I. Kukhar and E. S. Sivashova, *Physics of the Solid State*, **50** (2008) 1150.
[19] Bui Duc Hung, Nguyen Vu Nhan, Luong Van Tung and Nguyen Quang Bau, *Proc. Natl. Conf. Theor. Phys.* **37** (2012) 168.

The Hubbard–Stratonovich Transformation and Exchange Boson Contributions

Nguyen Van Hoa[1] **and Nguyen Tri Lan**[2]
[1]Pedagogical University, Hochiminh City, Vietnam
[2]Institute of Physics, Vietnam Academy of Science and Technology, 10 Dao Tan, Ba Dinh, Ha Noi, Vietnam

E-mail: nvhoa62@yahoo.com

Abstract. The Hubbard - Stratonovich transformation of many - body theory is reconsidered in various versions. Learning from its failures in coping with competing channels of collective phenomena, a combined Hubbard - Stratonovich transformation is proposed to overcome these difficulties. That yields a variety of fluctuating bosonic quantum fields describing the competition of different phases in the system under consideration. As a consequence of results, the validity of the Hubbard - Stratonovich transformation and its relation to a given order parameter are discussed.

1. Introduction

One of the most powerful techniques in the functional integral formalism is the use of the so - called Hubbard - Stratonovich transformation which is an exact mathematical transformation invented by Russian physicist R. L. Stratonovich [1] and popularized by British physicist J. Hubbard [2]. Since its birth, the Hubbard - Stratonovich transformation has a well - established place not only in many - body theory but also in elementary particle physics [3, 4]. It has led to a good understanding of important collective physical phenomena such as superconductivity, superfluidity of He^3, plasma and other charge - density waves, pion physics and chiral symmetry breaking in quark theories, etc. It has put heuristic calculations such as the Gorkov's derivation [5] of the Ginzburg - Landau equations [6] on a solid theoretical ground. In addition it is in spirit close to the famous density functional theory via the celebrated Hohenberg - Kohn and Kohn - Sham theorems. The transformation is cherished by theoreticians since it allows them to re-express a two - boby interaction exactly in terms of a bosonic collective field variable whose fluctuations, i.e. exchange bosons can in principle be described by higher loop diagrams. The only bitter pill is that any approximate treatment of a many - body system can describe interesting physics only if calculations may be restricted to a few low - order diagrams. This is precisely the point where the Hubbard - Stratonovich transformation fails. Trouble arises in all those many - body systems in which different exchange boson effects compete with similar strengths. Historically, an important example is the fermionic superfluid He^3. While BCS superconductivity was described easily via the Hubbard - Stratonovich transformation by transforming the two - body interaction to a field theory of Cooper pairs, this approach did initially not succeed in a liquid of He^3 atoms. Due to the strongly repulsive core of an atom, the forces in the attractive p - wave are not sufficient to bind the Cooper pairs. Only after taking the help of another exchange boson field that arises in the competing paramagnon channel into account, could the formation of weakly bound Cooper pairs be explained.

2. The Hubbard–Stratonovich Transformation

While the term of Hubbard - Stratonovich transformation may sound very impressive, the essence of this transformation is nothing but a straightforward manipulation of a Gaussian integral with respect to complex or Grassmann variables. In order to make this fact more transparent for a generic two - body interaction, let's briefly recall how it appears. Consider a non - relativistic many - body fermionic system described by a Hamiltonian:

$$H = H_0 + H_1, \tag{1}$$

where

$$H_0 = \sum_{\sigma}\sum_{\mathbf{k}} \varepsilon_\sigma(\mathbf{k})\, \psi_\sigma^\dagger(\mathbf{k})\, \psi_\sigma(\mathbf{k}), \tag{2}$$

H_0 is unperturbed Hamiltonian describing the system of free electrons, the symbols $\psi_\sigma(\mathbf{k})$ and $\psi_\sigma^\dagger(\mathbf{k})$ denote the annihilation and creation operators of the electrons with the spin projection $\sigma =\uparrow,\downarrow$ respectively, and $\varepsilon_\sigma(\mathbf{k})$ is the dispersion of the free electrons; and

$$\begin{aligned} H_1 \;=\; & \sum\nolimits_{\sigma_1,\sigma_2,\sigma_1',\sigma_2'} \sum\nolimits_{\mathbf{k},\mathbf{k}',\mathbf{q}} V_{\sigma_1\sigma_2\sigma_1'\sigma_2'}\left(\mathbf{k},\mathbf{k}',\mathbf{q}\right) \\ & \psi_{\sigma_1}^\dagger\left(\mathbf{k}-\tfrac{\mathbf{q}}{2}\right) \psi_{\sigma_2}^\dagger\left(\mathbf{k}'+\tfrac{\mathbf{q}}{2}\right) \psi_{\sigma_2'}\left(\mathbf{k}'-\tfrac{\mathbf{q}}{2}\right) \psi_{\sigma_1'}\left(\mathbf{k}+\tfrac{\mathbf{q}}{2}\right), \end{aligned} \tag{3}$$

H_1 is a generic effective two - body interaction term of interacting electrons written in the normal order, which may cover the contributions of other interactions in the system such as electron - phonon interaction, spin - spin interaction, ... even impurity - electron and impurity - impurity interactions. In both mathematical and physical aspects, the generic effective interaction term is very complicated and has no exact and explicit analytical expression.

The Hubbard - Stratonovich transformation enters the arena by rewriting the two - body interaction term (3) with the help of an auxiliary complex field $(\varphi)_{\sigma,\sigma'}(\mathbf{k},-\mathbf{q},\tau)$ as

$$\begin{aligned} & \exp\left\{ (-1)^\alpha \int_0^\beta \mathrm{d}\tau \sum\nolimits_{\sigma_1,\sigma_2,\sigma_1',\sigma_2'} \sum\nolimits_{\mathbf{k},\mathbf{k}',\mathbf{q}} \left(\psi^\dagger\psi\right)^\dagger_{\sigma_1'\sigma_1}(\mathbf{k},-\mathbf{q},\tau) \right. \\ & \qquad \left. V_{\sigma_1\sigma_2\sigma_1'\sigma_2'}\left(\mathbf{k},\mathbf{k}',\mathbf{q}\right) \left(\psi^\dagger\psi\right)_{\sigma_2\sigma_2'}\left(\mathbf{k}',-\mathbf{q},\tau\right) \right\} \\ & = \frac{1}{W}\int \left[\mathcal{D}\varphi^*\right]\left[\mathcal{D}\varphi\right] \exp\left\{ -S_0\left[\varphi^*,\varphi\right] \right. \\ & \left. -(i)^\alpha \int_0^\beta \mathrm{d}\tau \sum\nolimits_{\sigma,\sigma'} \sum\nolimits_{\mathbf{k},\mathbf{q}} \left\{ \begin{array}{l} (\varphi)^*_{\sigma'\sigma}(\mathbf{k},-\mathbf{q},\tau)\left(\psi^\dagger\psi\right)_{\sigma\sigma'}(\mathbf{k},-\mathbf{q},\tau) \\ +\left(\psi^\dagger\psi\right)^\dagger_{\sigma'\sigma}(\mathbf{k},-\mathbf{q},\tau)\,(\varphi)_{\sigma\sigma'}(\mathbf{k},-\mathbf{q},\tau) \end{array} \right\} \right\}, \end{aligned} \tag{4}$$

where

$$\begin{aligned} S_0\left[\varphi^*,\varphi\right] \;=\; & \int_0^\beta \mathrm{d}\tau \sum\nolimits_{\sigma_1,\sigma_2,\sigma_1',\sigma_2'} \sum\nolimits_{\mathbf{k},\mathbf{k}',\mathbf{q}} (\varphi)^*_{\sigma_2'\sigma_2}\left(\mathbf{k}',-\mathbf{q},\tau\right) \\ & (V)^{-1}_{\sigma_2'\sigma_1'\sigma_2\sigma_1}\left(\mathbf{k}',\mathbf{k},\mathbf{q}\right) (\varphi)_{\sigma_1\sigma_1'}(\mathbf{k},-\mathbf{q},\tau). \end{aligned} \tag{5}$$

The Hubbard - Stratonovich transformation is actually explicit and exact. However, in practice, to make the Hubbard - Stratonovich transformation a meaningful operation, it has to be motivated by many physical considerations. Normally, for bosonic Hubbard - Stratonovich

transformation, the interaction is split by choosing a composite operator consisted of two fermionic field operators which are called bifermionic.
The version of the transformation in (4) corresponds to the type of pairing which is referred to decoupling in the direct channel or in other words, in the "density" channel. However, there are, at least, two other inequivalent choices of pairing up the fermionic operators to construct the fermionic bilinear term of the generic two - body interaction which are decoupling in the "exchange" channel generated by choice of the fermionic pairings $(\psi^\dagger\psi)_{\sigma\sigma'}\left(\mathbf{k},\mathbf{k}',\mathbf{q}\right)$, and $(\psi^\dagger\psi)^\dagger_{\sigma'\sigma}\left(\mathbf{k}',\mathbf{k},\mathbf{q}\right)$, and decoupling the in Cooper channel corresponding to following bifermionic operators $(\psi\psi)_{\sigma_1\sigma_2}\left(\mathbf{k},\mathbf{k}',\mathbf{q}\right)$ and $(\psi^\dagger\psi^\dagger)_{\sigma_2\sigma_1}\left(\mathbf{k}',\mathbf{k},-\mathbf{q}\right)$. This is always not clear why one kind of pairing is chosen but not others.
The existence of there pairing possibilities may lead to the impression of a certain arbitrariness to the Hubbard - Stratonovich transformation philosophy. Indeed, the "right" choice of the decoupling field should be only motivated by physical reasoning. This is to emphasis that the transformation as such is always exact, no matter what channel is chosen.
Besides the physical reasoning of the system under consideration, the mathematical structure of the two - body interaction term has also an important role in choosing the "right" channel for the Hubbard - Stratonovich transformation. In the mathematical context, the "right" choice of the decoupling corresponds to the existence of the auxiliary bosonic Gaussian integral $\mathcal{W}$ (5), i. e. the positiveness and the existence of the inverse of two - body matrix element $(V)^{-1}_{\sigma_2'\sigma_1'\sigma_2\sigma_1}\left(\mathbf{k}',\mathbf{k},\mathbf{q}\right)$.
For every type of bifermionic operators as "density", "exchange" and "Cooper" pairings, the Hubbard - Stratonovich transformation requires a corresponding inverse of two - body matrix element such as

$$\sum_{\sigma,\sigma'}\sum_{\mathbf{k}'} V_{\sigma_1\sigma\sigma_1'\sigma'}\left(\mathbf{k},\mathbf{k}',\mathbf{q}\right)(V_{\rm d})^{-1}_{\sigma'\sigma_2'\sigma\sigma_2}\left(\mathbf{k}',\mathbf{k}_1,\mathbf{q}\right)=\delta_{\mathbf{k}\mathbf{k}_1}\delta_{\sigma_1\sigma_2}\delta_{\sigma_1'\sigma_2'}, \tag{6}$$

$$\sum_{\sigma,\sigma'} V_{\sigma_1\sigma\sigma'\sigma_1'}\left(\mathbf{k},\mathbf{k}',\mathbf{q}\right)(V_{\rm ex})^{-1}_{\sigma_2'\sigma'\sigma\sigma_2}\left(\mathbf{k}',\mathbf{k},\mathbf{q}\right)=\delta_{\sigma_1\sigma_2}\delta_{\sigma_1'\sigma_2'}, \tag{7}$$

$$\sum_{\sigma_1',\sigma_2'} V_{\sigma_3\sigma_4\sigma_1'\sigma_2'}\left(\mathbf{k},\mathbf{k}',\mathbf{q}\right)(V_{\rm C})^{-1}_{\sigma_2'\sigma_1'\sigma_2\sigma_1}\left(\mathbf{k}',\mathbf{k},\mathbf{q}\right)=\delta_{\sigma_1\sigma_3}\delta_{\sigma_2\sigma_4}, \tag{8}$$

respectively.
The existence of different inverses for the two - body matrix elements $V_{\sigma_1\sigma_2\sigma_1'\sigma_2'}\left(\mathbf{k},\mathbf{k}',\mathbf{q}\right)$ gives some additional arguments to physical insights of the system under consideration, deciding the existence of the corresponding order parameter.
In fact, in the most well - known microscopic models of interacting fermion systems, the two - body matrix element is dramatically simplified and its spin - dependence is actually neglected. Such way of simplification leads to the fact that all three possible decoupling schemes are always posible and can be applied to the interacting electron systems, or in other words, without spin - dependence of two - body matrix elements all possible kinds of order parameters can coexist in a theoretical calculations leading to "false" physical interpretation of the given system. And the only thing to justify physical properties of a real physical system described by these microscopic models is physical reasoning.
In order to verify this arguments, another version of the Hubbard - Stratonovich transformation

based on a simple and always valid bosonic Gaussian integral

$$\mathcal{W} = \int [\mathcal{D}\varphi^*] [\mathcal{D}\varphi] \exp\left\{ -\int_0^\beta d\tau \sum_{\sigma_1,\sigma_1'} \sum_{\mathbf{k},\mathbf{q}} (\varphi)^*_{\sigma_1'\sigma_1} (\mathbf{k}, -\mathbf{q}, \tau) (\varphi)_{\sigma_1\sigma_1'} (\mathbf{k}, -\mathbf{q}, \tau) \right\}, \tag{9}$$

is performed.

This version, however, requires the existence of term $(V)^{\frac{1}{2}}_{\sigma_1\sigma_2\sigma_1'\sigma_2'} \left(\mathbf{k}, \mathbf{k}', \mathbf{q}\right)$ and its inverse for each corresponding pairing.

Once the mathematical structure of the two - body interaction term confirms the possible order parameters which would appear in the systems, the standard techniques can be used to obtain physical quantities without difference among decoupling schemes.

The simplest physical reasoning to include spin–dependence to the two–body matrix element is the contributions of exchange bosons reflecting the interactions between conducting electrons and bosonic background fluctuations which cause the rise of selected competing channels. Mathematically, the exchange boson contributions of background fluctuations would lead to additional spin–dependent terms to two–body interaction, which allows the existence of specified version of the inverse two–body matrix element.

3. Discussions and Conclusions

The validity of the Hubbard - Stratonovich transformations in different decoupling schemes as "density", "exchange" and "Cooper" channels is discussed and shows that spin - dependence of two - body matrix elements plays an important role in the existence of a certain phase in the systems. The roles of exchange bosons in the spin–dependence of two–body matrix element are also discussed. This can motivate to attain a proper comprehension of formalism and that the functional integral formalism is a very systematic technique, free of ambiguity.

Acknowledgment

This work is supported by NAFOSTED Grant 103.02-2011.05.

References

[1] R. L. Stratonovich. On a method of calculating quantum distribution functions. *Soviet Physics Doklady*, 2:461, 1958.

[2] J. Hubbard. Calculation of partition functions. *Phys. Rev. Lett.*, 3(2):77–78, 1959.

[3] H. Kleinert. *Path Integrals in Quantum Mechanics, Statistics, and Polymer Physics.* World Scientific, Singapore, 3rd edition, 2004.

[4] John W. Negele and Henri Orland. *Quantum Many-Particle Systems*, volume 68 of *Frontiers in Physics*. Addison Wesley, 1988.

[5] L. P. Gor'kov. Microscopic Derivation of the Ginzburg-Landau Equations in the Theory of Superconductivity. *Sov. Phys. JETP*, 36:1364–1367, 1959.

[6] V. L. Ginzburg and L. D. Landau. To The Theory of Superconductivity. *Sov. Phys. JETP*, 20:1064, 1950.

Boltzmann–Gaussian Transition under Specific Noise Effect

Chu Thuy Anh, Nguyen Tri Lan and Nguyen Ai Viet

Institute of Physics, Vietnam Academy of Science and Technology, 10 Dao Tan, Ba Dinh, Ha Noi, Vietnam

E-mail: ctanh@iop.vast.ac.vn

Abstract. It is observed that a short time data set of market returns presents almost symmetric Boltzmann distribution whereas a long time data set tends to show a Gaussian distribution. To understand this universal phenomenon, many hypotheses which are spreading in a wide range of interdisciplinary research were proposed. In current work, the effects of background fluctuations on symmetric Boltzmann distribution is investigated. The numerical calculation is performed to show that the Gaussian noise may cause the transition from initial Boltzmann distribution to Gaussian one. The obtained results would reflect non–dynamic nature of the transition under consideration.

1. Introduction

Market returns have received much attention and have been studied for decades[7, 8, 9, 10, 3, 5]. Empirical data has been used to illustrate some dynamic models of market[4]. Gaussian distribution seems to be the most used approximation to describe real–valued random variables which tend to cluster around a single mean value [1, 11]. Theoretically, Gaussian distribution could be used to model market returns distribution when studying the market takes for long time enough, i.e. when the returns set is big enough, not for the not–big–enough data set, or for some assets that can not be described dynamically.

The model proposed in this paper is another view without taking into account the time role, but the sets of background fluctuations. In the ideal case, just as perfect condition in a physical system, there is no background fluctuation. However, for all real market data, there always are various parameters affecting more or less on each movement of the market, such as political situation in a country, or disaster, or information disturbance, ...etc. These parameters could be considered as background fluctuations which will be describe as a set of random auxiliary variables. By including the background fluctuations into the data set, market returns tendency distribution can be studied independently to time value.

2. Standard Distributions

Conventionally, the return obtained from the long time data set is modeled by a Gaussian distribution, and the return from the short time data set is empirically presented by a symmetric Boltzmann distribution. As the initial stage these standard distributions are recalled.

Consider an asset with random variables in an ideal condition, means without background fluctuations, then probability distribution of returns reduces to exponential and Gaussian[6, 2].

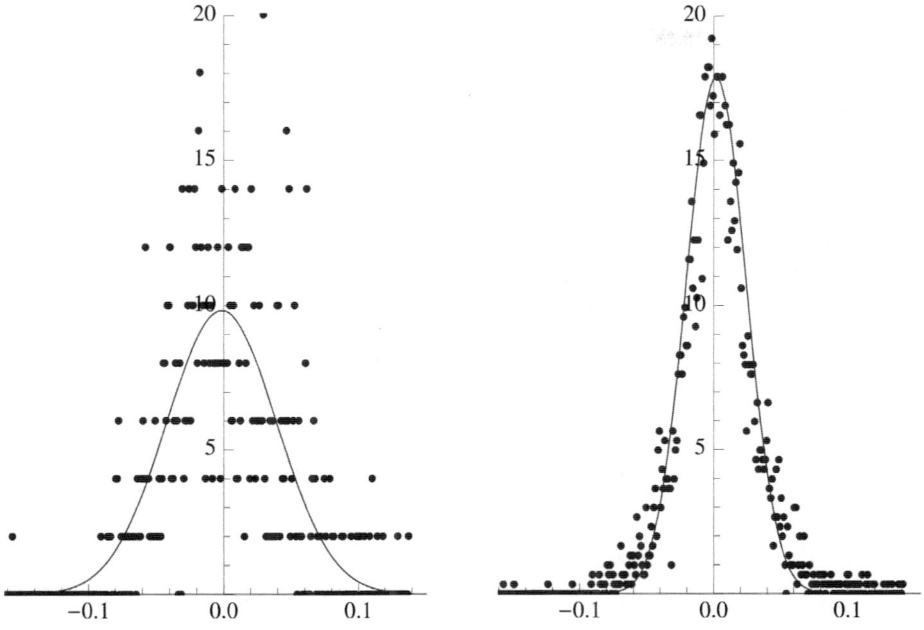

Figure 1. The left graphic presenting data set of return of Apple Inc. in time period 2000 − 2002 is not able to be well fitted with a Gaussian distribution, whereas the right one presenting data set of return of the same stock in time period 2000 − 2012 is more fitted with a Gaussian distribution. Longer time of data set, the graphics is more well fitted with the Gaussian one.

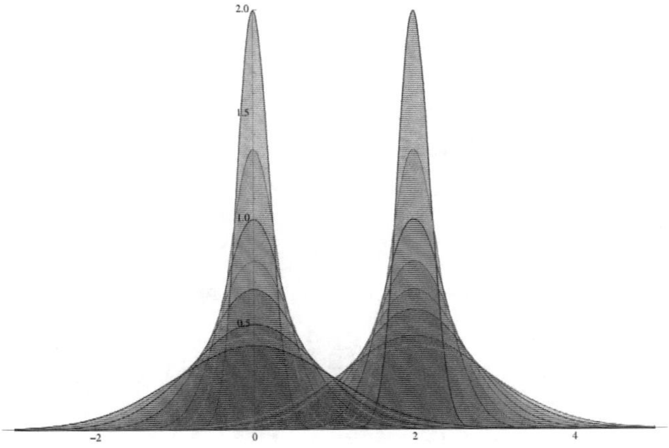

Figure 2. Zero and Non–Zero Gaussian distributions with different values of variance σ

A normal distribution is written as

$$f_G(\sigma, x_0, x) = \frac{1}{\sqrt{2\pi}\sigma} e^{-\frac{(x-x_0)^2}{2\sigma^2}} \tag{1}$$

where x_0 and σ are mean value and variance, respectively.
In other hand, a symmetric Boltzmann distribution is written under the form

$$f_B(\lambda, x_0, x) = \frac{1}{2}\lambda e^{-\lambda|x-x_0|} \tag{2}$$

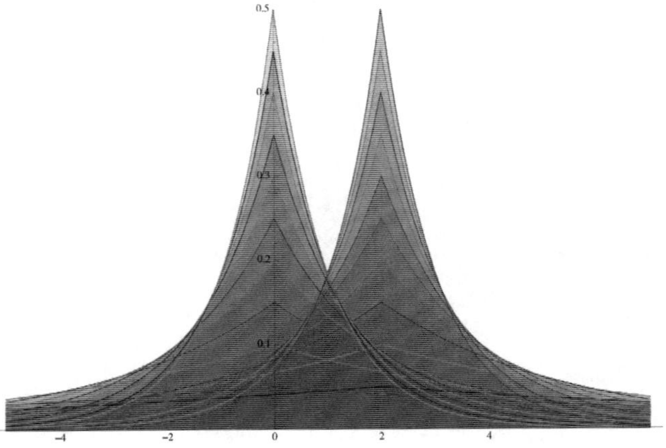

Figure 3. Zero and Non–Zero Boltzmann distributions with different values of parameter λ

where λ plays the role of inverse of market temperature, and x_0 is central value of given distribution.

3. The contributions of background fluctuations

In most of the cases, a random variable is not absolutely independently random. In fact, the random variables are all affected by a set of other random variables that could be considered as background or environmental fluctuations, which should be described by a stochastic process. Using the same physical approach as in relativistic energy expressions, random quantities will be changed under background fluctuations just the same way as thermal fluctuations in thermodynamic systems, and it can be described

$$x^2 \to x^2 + \sum_i \epsilon_i^2. \tag{3}$$

For the sake of simplicity, the current consideration is limited in the case of $i = 1$. Hence, the distribution functions would be rewritten as

$$f(x) \to C(\epsilon) f\left(\sqrt{x^2+\epsilon^2}\right), \tag{4}$$

in which $C(\epsilon)$ is the normalization constant of the distribution function $f\left(\sqrt{x^2+\epsilon^2}\right)$, i.e.

$$C(\epsilon) \int f\left(\sqrt{x^2+\epsilon^2}\right) dx = 1. \tag{5}$$

It has been mentioned by Fokker–Plank function, under the influence of drag forces and random forces, the probability density function of a particle's velocity is described dynamically, as in Brownian motion, as well as all in this case. This is a dynamic description, in which the distribution tends to Gaussian one when time closes to infinity, means on this interval of time, there are enough fluctuations to make a tends–to–Gauss transition.

A close–to–infinity time series could be seen as a collection of fluctuations which is big enough to make a distribution to be a tends–to–Gauss one. The proposed model in this paper does not depend on a specific dynamic, and somehow a possibility to define relaxation time between two almost–balance–states is supplied. The proposed model can be used to describe not only equilibrium–equilibrium process but also non–equilibrium to equilibrium process.

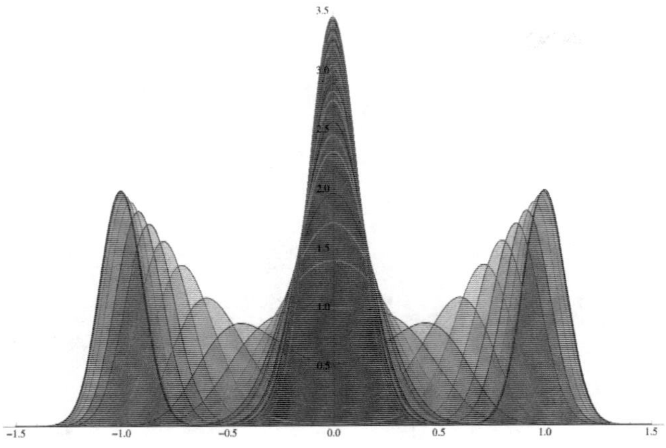

Figure 4. In the present of small noise, Gaussian distribution splits itself into two–pick distributions, and two picks get closer in the regime of large noise.

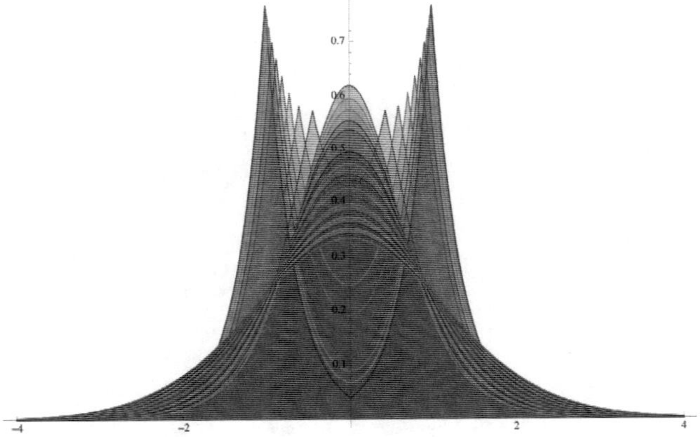

Figure 5. The same phenomenon happens with the Boltzmann distribution in present of various values of noise.

There are two types of symmetric distribution under consideration, zero centered distribution and non–zero centered one. Under background fluctuation effect, the non–zero centered distribution split into two symmetric distributions, whereas the zero centered one doesn't split into two distributions but degenerated and collapsed to a more stable Gaussian distribution. This phenomenon is quite the same as the picture of particle and antiparticle in a physics system. The zero–centered case could be treated as a physics system in an ideal condition, vacuum one. In the present of noise, the Gaussian and Boltzmann distributions split into two–pick distribution in the small value regime of noise, and bold single pick distribution in the large value regime of noise.

The transition of a symmetric or quasi–symmetric distributions to Gaussian distribution is a general phenomenon, which does not depend on a specific dynamic of noise, but depends on distribution of fluctuation only. The parameters of background fluctuations such as mean value and variation lead to the change of corresponding parameters of final Gaussian distribution. However, in real data set, it is impossible to directly observe and to classify the background fluctuations, their contributions are silently included into each element of data set. It can only be observed the transition of the initial distribution to final one under influence of background

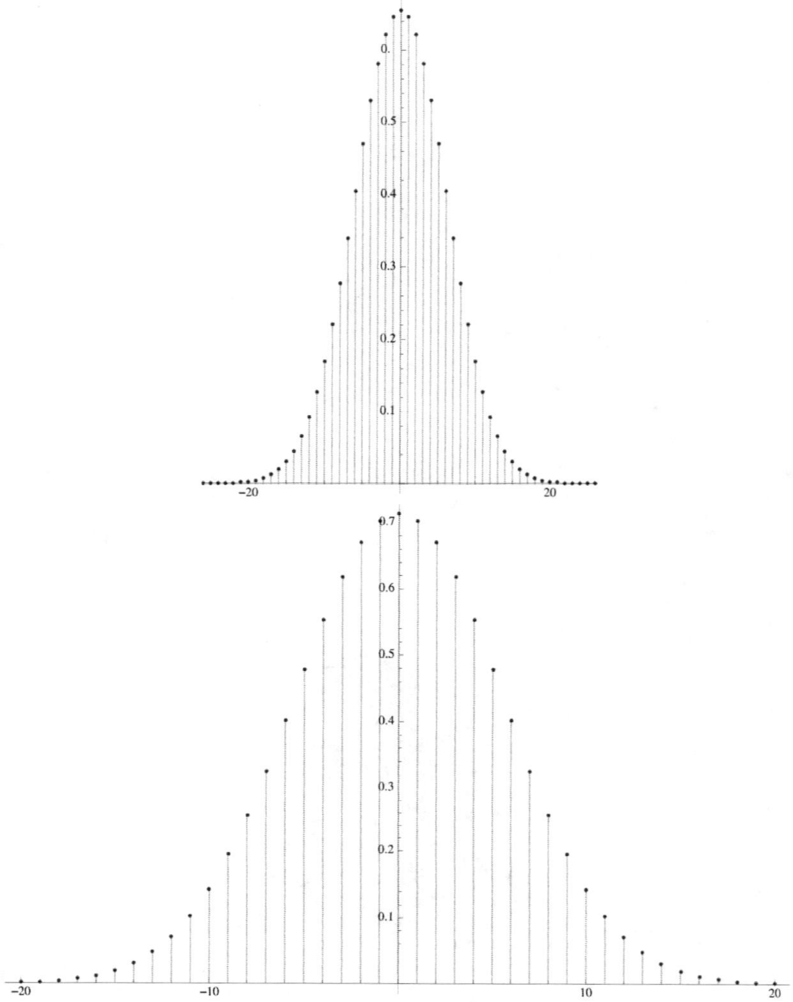

Figure 6. The numerical summation over full range of Gaussian noise presents the final Gaussian distribution which does not depend on initial distribution.

fluctuations. Mathematically, the final distribution $f_0(x)$, which is obtained from real data set, is the sum of all possible contributions of background fluctuations with specific statistical weight $f^G(\epsilon)$, i.e.

$$f_0(x) \to \int f^G(\epsilon) C(\epsilon) f(\sqrt{x^2+\epsilon^2}) d\epsilon. \qquad (6)$$

Initial Boltzmann distribution affected by all possible background fluctuations then tends to Gaussian distribution. It can be shown analytically by taking integral of probability distribution function for all noises.

In general cases background fluctuations follow Gaussian law. Empirical data shows that under background fluctuations, most of the distributions tend to Gaussian-liked ones before transforming to Gaussian one. As an inverse approach, by analyzing the long time data set the role of background fluctuations and their statistical parameters can be emerged and determined, and the obtained results provide a quantitative insight of background fluctuations in many phenomena of physics and interdisciplinary physics.

Journal of Physics: Conference Series **537** (2014) 012005 doi:10.1088/1742-6596/537/1/012005

The condition where it is supposed to exist the Gaussian–distribution–form fluctuation is a kind of ideal one. In the case of econophysics assets, the data has been taken in specific time interval, means with incomplete Gaussian distribution fluctuations. Background fluctuations drive the initial distribution to a Levy distribution. This problem will be discussed numerically and analytically in next papers. In the context of this paper the Levy part is cut off.

4. Conclusion

A non–dynamic approach has been proposed, taking into account the important role of background fluctuations. Background fluctuations in an economic system play the same role as in a physics one. Normal distribution fluctuations guide the transition process from exponential distribution to normal distribution. The collapse process has been draw numerically and also analytically in the paper.

Taking this result into economic view, it means no matter how many fluctuations it might have for an economic object, if all are random fluctuation, the probability distribution always tends to Gaussian one. If not, that mean some fluctuations are not considered yet.

For some special cases that background fluctuation does not follow Gaussian law, the probability distribution will be rearranged differently. Theses cases will be treated in our next works.

References

[1] Victor M. Yakovenko A. Christian Silva, Richard E. Prange. Exponential distribution of financial returns at mesoscopic time lags: a new stylized fact. *Physica A*, (344), 2004.

[2] V. M. Yakovenko A. Dr. *Quant. Finance 2*, (443), 2002.

[3] V.M. Yakovenko A.C. Silva. *Physica A*, 324(303), 2003.

[4] Thuy Anh Chu, Hong Lien Do, and Ai Viet Nguyen. Simple model for market returns distribution. *Communications in Physics*, 23(2), 2013.

[5] Stauffer D. Phys. rev. e 60 (1999) 6519. *Physica A*, (336):1–5, 2004.

[6] A. Silva et al. *Physica A 344*, (344):277–235, 2004.

[7] M. Potters J.-P. Bouchaud. Theory of financial risks. *Cambridge University Press, Cambridge*, 2000.

[8] H. Salunkay H.E. Stanley K. Matia, M. Pal. *Europhys. Lett.*, 66(909), 2004.

[9] R. Riera L.C. Miranda. *Physica A*, 297(509), 2001.

[10] L.N. Amaral M. Meyer H.E. Stanley V. Plerou, P. Gopikrishnan. *Phys. Rev. E*, 60(6519), 1999.

[11] V. M. Yakovenko. Econophysics, statistical mechanics approach to. In *Encyclopedia of Complexity and System Science*. Springer, 2007.

Journal of Physics: Conference Series **537** (2014) 012006 doi:10.1088/1742-6596/537/1/012006

The Two–Level Model on Absorption Spectra of Oxyhemo- and Neuroglobin

To Thi Thao[1], Sophie Bernad[2], Valerie Derrien[2], Chu Thuy Anh[1], Nguyen Tri Lan[1] and Nguyen Ai Viet[1]

[1]Institute of Physics, Vietnam Academy of Science and Technology, 10 Dao Tan, Ba Dinh, Ha Noi, Vietnam
[2]University of Paris Sud, Paris, France

E-mail: ctanh@iop.vast.ac.vn

Abstract. The absorption spectra of deoxyhemoglobin and oxyneuroglobin has been examined. The well–known two–level model has been used to study the absorption spectra. The model has been verified by experemental data for heomoglobin and neuroglobin, that gives good fittings. The discussion on the obtained results leads to further works

1. Introduction

The globin family evolved from a common ancestor. Its classical representatives, hemoglobin (Hb) and myoglobin (Mb), have been widely investigated for more than four decades[2], while neuroglobin was just identified in the nervous system of vertebrates[1] for more than ten years. Globin family has been studied for many years under different ways of view. Among which absorption spectra is one of the most effective method to touch various physical and chemical aspects of the members of globin family. Different physiological responses can be emphasized by imaging at different wavelengths[7]. Globins are heme–containing proteins involved in binding and/or transporting oxygen. There exist several stable states of globins characterized by the oxidation states of the heme iron, the heme orientation and the nature of the sixth ligand of the iron[3]. Oxygen binds in an "end–on bent" geometry where one oxygen atom binds Fe and the other protrudes at an angle. When oxygen is not bound, a very weakly bonded water molecule fills the site, forming a distorted octahedron. In binding, oxygen temporarily and reversibly oxidizes (Fe^{2+}) to (Fe^{3+}) while oxygen temporarily turns into superoxide, thus iron must exist in the +2 oxidation state to bind oxygen. If superoxide ion associated to Fe^{3+} is protonated the hemoglobin iron will remain oxidized and incapable of binding oxygen[9]. It has been observed[10][4] that the visible absorption spectrum of deoxyhemoglobin has one pick, while the visible absorption spectra of oxyhemoglobin has two picks. It is also observed that for a sample of oxygenated neuroglobin, the distribution has only one pick. The phenomenon seems to contradict to the observation on hemoglobin. But theoretically it could be modeled by a two–state system, which solves the problem and draws a line through out the absorption spectra of globin family.

Journal of Physics: Conference Series **537** (2014) 012006 doi:10.1088/1742-6596/537/1/012006

In theory, the two–level model is widely applied to consider a number of physical and chemical systems, such as spin$-1/2$ particles are two–state quantum systems when only the spin degree of freedom is considered. It also used to study the "inverting" degree of freedom in an ammonia molecule; the nitrogen at the vertex of an ammonia molecule exhibits two molecular states – "up" and "down", corresponding to opposite positions with respect to the plane of the three hydrogen atoms. In an electric field, these two states are non-degenerate.

This work uses two–level model in studying globin group, builds a model which works well with the absorption spectrum of hemoglobin.

2. Materials and methods

The deoxyhemoglobin and oxyneuroglobin samples have been investigated, using two–level model. A sample of bovin Hgb was prepared, absorbance measurements were made at room temperature (20–24$°C$) in the spectral range of 450–700nm. A sample contains pure 100% Ngb and hemin was examined in the spectral range of $280 - 700nm$, in different conditions of $pH = 2.3, 5.1, \text{and}\, 8.1$.

In general, the absorption intensity varies in function of the wavelength. It has been already indicated[8] that absorption intensity of deoxyhemoglobin is proportional to the wavelength when wavelength approaches to $555nm$ and inversely proportional to the wavelength when it is longer than $555nm$. In the case of oxyhemglobin, the absorption intensity reaches its maxima at $\lambda_1 \approxeq 540nm$ and $\lambda_2 \approxeq 580nm$. The energy diagram of hemogblobin has been illustrated in Fig. 1..

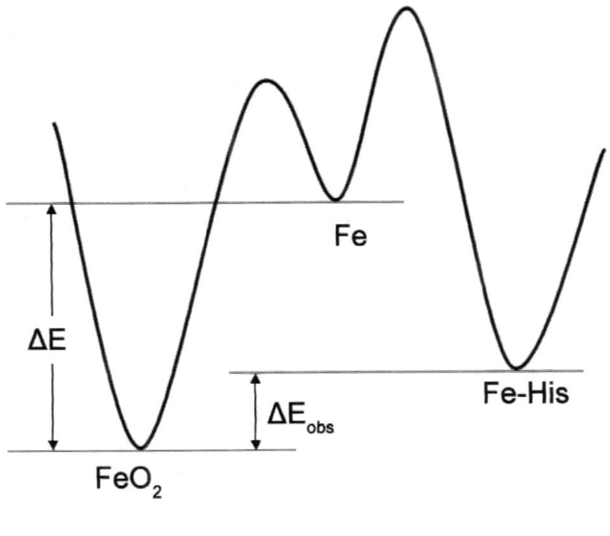

Figure 1. *Energy diagram of hemoglobin group*

The energy diagram of hemogobin group leads to an idea of using the two–level model to describe physics of heme, in which the protein would be considered as a system with only two quantum

Journal of Physics: Conference Series **537** (2014) 012006 doi:10.1088/1742-6596/537/1/012006

states [5, 6]. At the relaxed state, the system has only one pick, at excited state, the system has two pick. It is clear that with the experimental data[11], the distribution of deoxyhemoglobin absorption intensity has the gaussian–like shape. The two–level description would be useful to investigate the protein and to compare the theoretical results with this experimental data set.

2.1. An introduction to the two–level model

The two–level system which is also known as two–state system (TLS), is the simplest quantum system that can exist. Two–state formalism can be used to describe simple mixing of states which leads to phenomenon such as resonance stabilization and other level crossing related symmetries, so that it has a wide variety of application in physics, chemistry and biology.

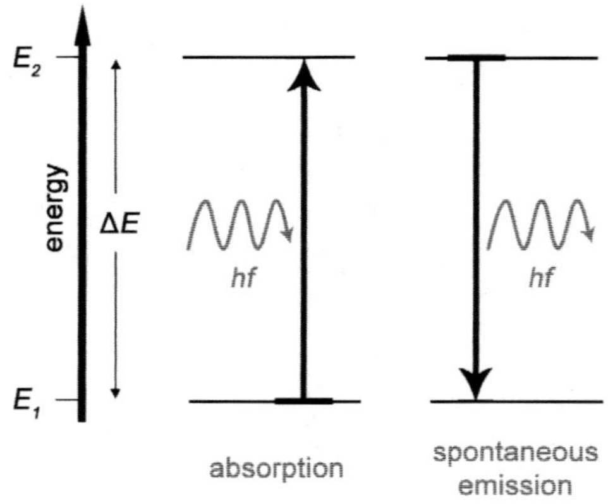

Figure 2. *Energy diagram of two-state quantum system*

The most well–known application of two–state system is of the spin of a spin $1/2$ particle such as an electron, whose spin can take value $+\hbar/2$ or $-\hbar/2$, where $\hbar$ is the reduced Plank constant. Another example, frequently studied in atomic physics, is the transition of an atom to or from an excited state, where the TLS formalism is used to quantitatively explain stimulated and spontaneous emission of photons from excited atoms.

The Hilbert space describing such a system fills up with only two degrees of freedom. Therefore, a complete basis spanning the space will consist of two independent states. The physics of a quantum mechanical two–state system is trivial if both states are degenerate. That is, if the states have the same energy. However, if there is an energy difference between the two states, then nontrivial dynamics can occur.

2.2. Dynamics for a time–independent Hamiltonian of the two–level system

Let H denote the time–independent Hamiltonian, and let $|\psi_1\rangle$ and $|\psi_2\rangle$ denote the two energy eigenstates of the system, with respective eigenvalues E_1 and E_2. Any state $|\psi(t)\rangle$ of the two–level system can be written as a superposition of the given energy eigenstates. In particular, at time $t = 0$, the state $|\psi(0)\rangle$ would read

$$|\psi(0)\rangle = c_1 |\psi_1\rangle + c_2 |\psi_2\rangle , \qquad (1)$$

Journal of Physics: Conference Series **537** (2014) 012006 doi:10.1088/1742-6596/537/1/012006

where c_1 and c_2 are complex coefficients with normalized condition $|c_1|^2 + |c_2|^2 = 1$.

At some later time t, the energy eigenstates $|\psi_1\rangle$ and $|\psi_2\rangle$ will have evolved into $|\psi_1\rangle e^{-iE_1 t/\hbar}$ and $|\psi_2\rangle e^{-iE_2 t/\hbar}$, respectively, and so

$$|\psi(t)\rangle = c_1 |\psi_1\rangle e^{-iE_1 t/\hbar} + c_2 |\psi_2\rangle e^{-iE_2 t/\hbar}. \tag{2}$$

Every two–level system has a characteristic angular frequency given by

$$\omega = \frac{E_2 - E_1}{\hbar}, \tag{3}$$

where it has been assumed that $E_2 > E_1$.

The physics of two–state systems can be usefully applied to multi–state systems where the system is known to have only enough energy available to excite the lowest two states, thus effectively creating a two–state system. In fact, in the nature, it is difficult to identify any true two–state systems, merely systems where the energetics of the circumstances isolate two particular states.

In other way, the set of all possible states in a two–level system can be mapped onto a representation known as the Bloch sphere (Fig. 3.).

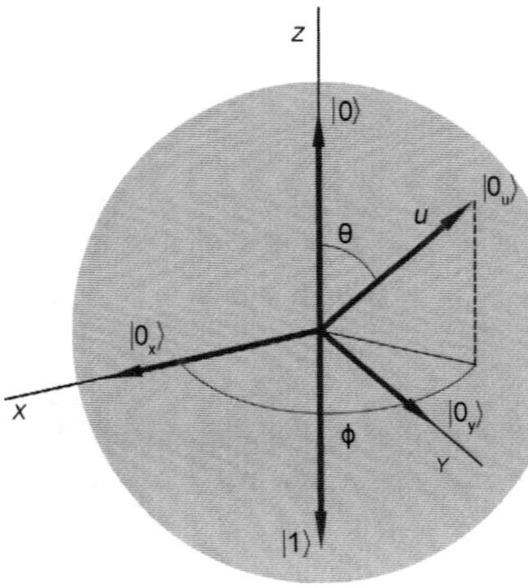

Figure 3. *Bloch's sphere for all possible states of a two–state system*

In this formalism, each state is represented as a point on the unit sphere. The spherical coordinates are usually chosen so that the energy eigenstate $|\psi_1\rangle$ has colatitude $\vartheta = 0$ and the energy eigenstate $|\psi_2\rangle$ has $\vartheta = \pi$, that is, $|\psi_1\rangle$ sits at the north pole and $|\psi_2\rangle$ at the south pole. A general state $|\psi\rangle$ has colatitude ϑ and azimuth φ determined by the coefficients c_1 and c_2.

2.3. Two–level system approach to hemoglobin group

Several experiments on globin family have showed that globins are heme–containing proteins involved in binding and/or transporting oxygen. The stable states of globins are characterized by the oxidation states of the heme iron as illustrated in Fig. 4., the heme orientation and the nature of the sixth ligand of the iron[3]. It has been observed[10][4] that the visible absorption spectrum of deoxyhemoglobin has one pick, while the visible absorption spectra of oxyhemoglobin has two picks. However, for a sample of oxygenated neuroglobin, the spectra has only single pick. The phenomenon seems to contradict to the observation on hemoglobin. But theoretically it could be modeled by a two–state system, which solves the problem and draws a line through out the absorption spectra of globin family.

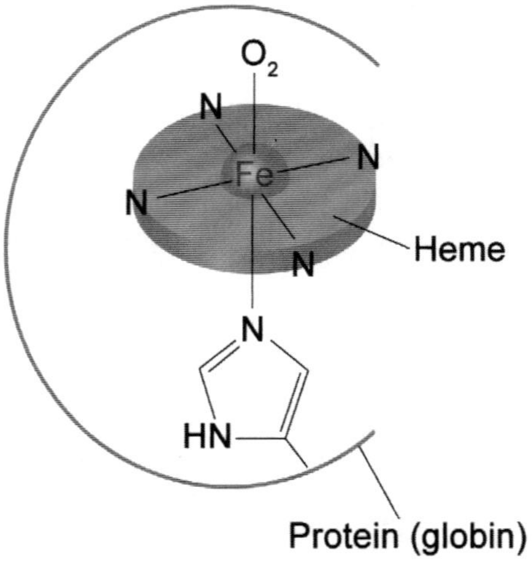

Figure 4. *Heme element in hemoglobin*

In the language of second quantization, the Hamiltonian of a two–state system reads

$$H = E_1 a^\dagger a + (E_2 + i\Gamma)\, b^\dagger b + \gamma \left(a^\dagger b + b^\dagger a\right), \tag{4}$$

where $a^\dagger$, a and $b^\dagger$, b are the creation and annihilation operators on the ground and excited states, $|\psi_1\rangle$ and $|\psi_2\rangle$, with energies E_1 and E_2, respectively, and γ is coupling coefficient between ground and excited states, which must be exponentially proportional to energy difference of two given states

$$\gamma \approx e^{-\beta(E_2 - E_1)}, \tag{5}$$

with $\beta = \frac{1}{k_B T}$ denotes the inverse of temperature.

Following standard analytical procedure which can be easily found in textbooks of quantum field theory, the cross-section of proposed model would be determined by the expression

$$\frac{d\sigma}{d\Omega} \simeq |M| \tag{6}$$

$$\approx \delta\left(E_2 - E_1 + \hbar\omega + i\Gamma\right). \tag{7}$$

Since it is well–known that

$$\delta\left(E_2 - E_1 + \hbar\omega\right) = \lim_{\Gamma \to 0} \delta\left(E_2 - E_1 + \hbar\omega + i\Gamma\right), \tag{8}$$

Journal of Physics: Conference Series **537** (2014) 012006 doi:10.1088/1742-6596/537/1/012006

and

$$\lim_{\Gamma\to 0} \exp\left\{-\frac{E^2}{\Gamma^2}\right\} = \delta(E). \tag{9}$$

The last expression enlightens the signification of parameter Γ, the width of the Gaussian function.

In this work, firstly we examine the proposed model for the absorption of deoxyhemoglobin, or in other words, we are going to determine the coupling coefficient Γ from the absorption spectra of deoxyhemoglobin.

3. Numerical calculation and results

The Gaussian function is used

$$f(x) = \frac{a}{\sqrt{2\pi}\Gamma} \exp\left\{-\frac{(x-\mu)^2}{\Gamma^2}\right\}, \tag{10}$$

to find the width Γ of the pick in the absorption spectra of deoxyhemoglobin.

For the deoxyhemoglobin sample, the deoxidized hemoglobin ($Fe(II)$) is considered as ground state. In binding, oxygen temporarily and reversibly oxidizes (Fe^{2+}) to (Fe^{3+}) and the hemoglobin goes to excited state. The absorption spectra of deoxidized hemoglobin corresponds to the relax line in energy diagram.

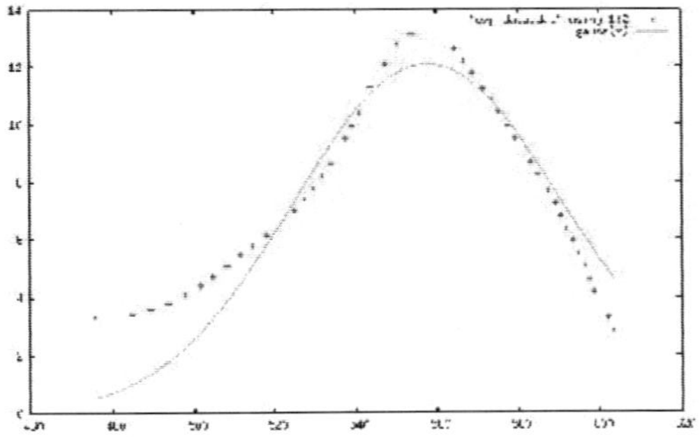

Figure 5. *Absorption intensity of deoxyhemoglobin sample. The measurements have been done in the range of wavelength from 450nm to 700nm. Numerical calculation with ratio 1/2: $\Gamma = 32.9$; $a = 997.5$ and $\mu = 587$*

The numerical calculation for oxidized hemin of neuroglobin sample at different values of pH provides the clues of the influence of pH on the width parameter Γ. It seems that the width parameter depends strongly on the inverse of pH. The dependence of the width in the absorption spectra on pH as a function $\Gamma(x_{pH})$ would be considered in a future works.

Moreover, absorption spectrum of neuroglobin at low pH presents only one pick and it tends to split in two picks while pH is increased. The observed spectra form of oxidized neuroglobin seems to be the same as for deoxihemoglobin for the very low pH, means for the acid environment. In acid environment, the protein has a strong binding with environment, the observed spectra could be considered as for the hemin alone. While increasing the pH of the environment, the binding

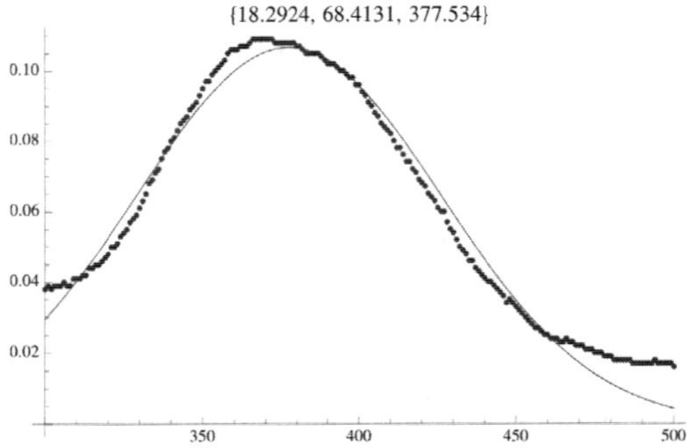

Figure 6. *Absorption intensity of Ngb and hemin sample. Wavelength varies from 280 to 700 nm.* $pH = 2.3$. *Numerical calculation with* $\Gamma = 18.292$; $a = 68.413$ *and* $\mu = 377.534$

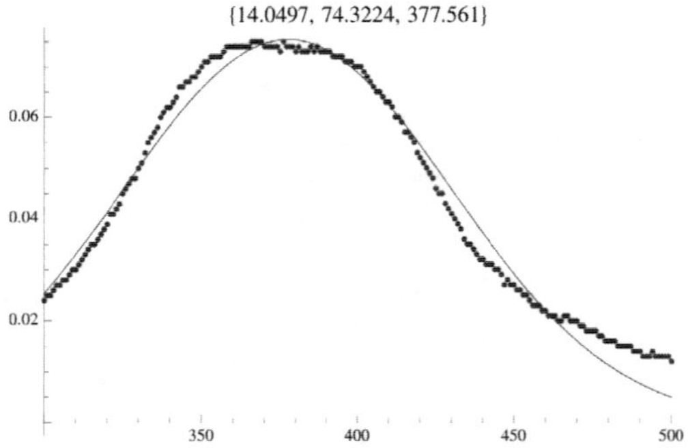

Figure 7. *Absorption intensity of Ngb and hemin sample. Wavelength varies from 280 to 700 nm,* $pH = 5.1$. *Numerical calculation with* $\Gamma = 14.049$; $a = 74.322$ *and* $\mu = 377.561$

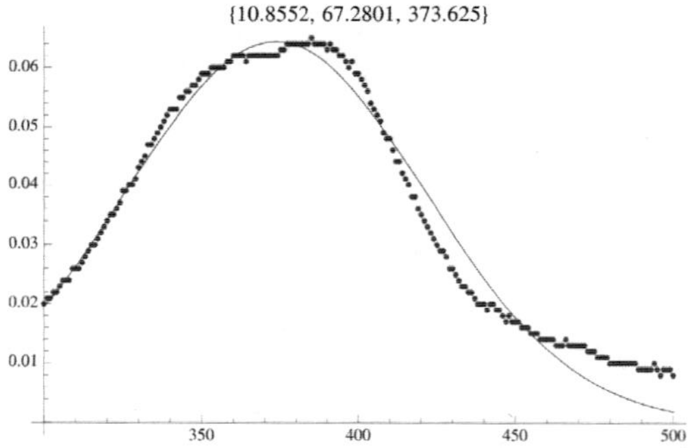

Figure 8. *Absorption intensity of Ngb and hemin sample. Wavelength varies from 280 to 700 nm,* $pH = 8.1$. *Numerical calculation with* $\Gamma = 10.855$; $a = 67.280$ *and* $\mu = 373.625$

between heme and protein becomes stronger, the spectra is affected by a large noise caused by the heme-protein binding.

In more general point of view, the absorption spectrum of oxidized neuroglobin could be considered as degenerated spectrum, in which two picks reach almost the same value, at which two resonance wavelengths quite close one to another, the change of environmental pH has an effect of external field causing splitting degenerated spectrum into two separated pick spectra.

The difference between hemoglobin and neuroglobin absorbtion spectrum is caused by the difference of their symmetry.

4. Discussion

A model has been built for absorption spectrum of deoxihemoglobin and oxidized neuroglobin. The model gives a good agreement with experiment data, from which some parameters could be determined. A qualitative explanation of some hemin's behaviors is reasoned in the context of the two–level model.

The absorption spectrum of deoxihemoglobin is well fitted with the Gaussian distribution function, while the absorption spectrum of oxydized neuroglobin seems to be the superposition of two spectrum corresponding to two oxygen–iron binding, which reaches almost the same maximum value. This hypothesis will be the main subject of next work.

The model will be optimized and some other parameters will be determined in the next work, where further knowledge of the behaviors of oxihemoglobin and oxidized neuroglobin is presented.

Acknowledgement

This work is supported by NAFOSTED Grant 103.06-2011.51 and 103.02-2011.05.

References

[1] Alessio Bocedi1 Paolo Ascenzi1 Sylvia Dewilde2 Pesce A Bolognesi M Bocedi A Ascenzi P Dewilde S Moens L Hankeln T Burmester T. Alessandra Pesce, Martino Bolognesi+. Neuroglobin and cytoglobin. fresh blood for the vertebrate globin family. *EMBO Rep.*, 3(12):1146–1151, 12 2002.

[2] Pierre Sebban Anthony Bocahut, Sophie Bernad and Sophie Sacquin-Mora. Relating the diffusion of small ligands in human neuroglobin to its structural and mechanical properties. *J. Phys. Chem. B*, 113(50):16257–16267, 2009.

[3] Bernad S Sebban P Sacquin-Mora S Guittet E Lescop E. Bocahut A, Derrien V. Heme orientation modulates histidine dissociation and ligand binding kinetics in the hexacoordinated human neuroglobin. *J Biol Inorg Chem.*, 18(1):111–122, 2013.

[4] Yiting Cao Mark W. Dewhirst Brian S. Sorg ; Benjamin J. Moeller, Owen Donovan. Hyperspectral imaging of hemoglobin saturation in tumor microvasculature and tumor hypoxia development. *J. Biomed. Opt.*, 10(4), 8 2005.

[5] Griffiths J. David. *Introduction to Quantum Mechanics (2nd ed.)*. Pearson Prentice Hall, 2004.

[6] R.P. Feynman. *The Feynman Lectures on Physics: Volume 3.* Addison Wesley, 1965.

[7] Neil A Martin Arthur W Toga Nader Pouratian, Andrew F Cannestra. Intraoperative optical intrinsic signal imaging: a clinical tool for functional brain mapping. *Neurosurgical FOCUS*, 11 2002.

[8] Matthew Pincus Richard McPherson. *Henry's Clinical Diagnosis and Management by Laboratory Methods.* Elsevier, 2011.

[9] Charles D. Conover Rita Linberg and Kwok L. Shum. Hemoglobin based oxygen carriers: How much methemoglobin is too much? *Artificial Cells, Blood Substitutes and Biotechnology*, 26(2):133–148, 1998.

[10] Philip A. Wright Tom Lister and Paul H. Chappell. Optical properties of human skin. *J. Biomed. Opt.*, 17(9), 9 2012.

[11] A. Buursma W. G. Zijlstra. Spectrophotometry of hemoglobin: Absorption spectra of bovine oxyhemoglobin, deoxyhemoglobin, carboxyhemoglobin, and methemoglobin. *Comp. Biochem. Physiol.*, 118B(4):743–749, 1997.

Journal of Physics: Conference Series **537** (2014) 012007 doi:10.1088/1742-6596/537/1/012007

Two-Dimensional Electron Gas at the N-type Interface of the $LaAlO_3/SrTiO_3$ Hetero–Structure: A First Principle Study

N T Trang, T V Nam, P B Duy and G H Bach

Computational Material Science Laboratory, Faculty of Physics, Hanoi University of Science, 334 Nguyen Trai, Thanh Xuan, Hanoi, Vietnam

E-mail: trangnguyenphys@gmail.com

Abstract. We examine the formation of the two-dimensional electron gas (2-DEG) at the n-type interface of two band insulators $LaAlO_3$ (LAO) and $SrTiO_3$ (STO) by increasing the thickness of the LAO layers in the framework of the density functional theory (DFT). It has been shown that a metal-insulator phase transition occurs in the 2-DEG when the LAO layers reach to a critical thickness $d_c = 4$ monolayers. The increase of the LAO layers enhances the LAO potential so that electrons can transfer from the LAO valence band to the STO conduction band, which confines conducting electrons only in the STO substrate.

1. Introduction

Recently, a new interest in the hetero-structure between the two typical perovskites, $LaAlO_3$ (LAO) and $SrTiO_3$ (STO), has been emerged due to the observation of its extraordinary behaviors. Both of the mother compounds are conventional band insulators and non-magnetic. However, their hetero-interface exhibits a two-dimensional (2D) conductivity [1] with a metal-insulator (M-I) phase transition at a critical thickness of the LAO layers [2]. Moreover, the coexistence of superconductivity and magnetic phase like magneto-resistance, magnetic hysteresis have been found in this hetero-structure [3, 4, 5, 6], whose mechanism has not been fully understood yet. It should be noted that these phenomena were obtained only for the (001) n-type LAO/STO interface where the LAO terminates at the LaO atomic layer and the STO terminates at the TiO_2 atomic layer (see Fig. 1).

Concerning the transport properties of the n-type (001) LAO/STO hetero-structure, though several carrier sources, e.g. the polar discontinuity at the interface, oxygen vacancy defect (O_V), atomic intermixing at the interface, were proposed to be responsible for the conductivity of the hetero-structure [7, 8, 9], the 2D conductivity is believed to originate from the interface polar discontinuity. The simplest model basing on the polar discontinuity ideal, polar catastrophe (PC) model, argues that since neighboring atomic layers in (001) LAO slab are oppositely charged, i.e. $(LaO)^{+1}$ and $(AlO_2)^{-1}$, the LAO slab is polarized. At the same time, atomic layers in the (001) STO slab are neutral so that it is non-polarized. The discontinuity of the polar displacement at the interface gives rise to the divergence of the electrostatic potential within the LAO slab, which is energetically non-optimal. Therefore, a half of electron per unit cell (u.c) would be transferred from the LAO to the STO to avoid the potential divergence.

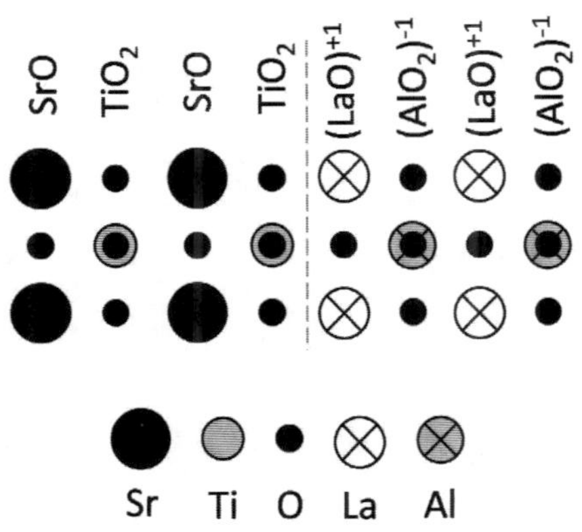

Figure 1. The n-type LAO/STO interface along (001) direction. The dash line denotes the interface.

This transferred charge forms a 2D electron gas (2-DEG) with the PC-predicted sheet carrier density $n_s^{PC} = 0.5\,e/u.c = 3.3 \times 10^{14}\,e/cm^{-2}$ around the interface which is responsible for the 2D conducting behavior of the system. Because of the simplicity, the PC model has some drawbacks. Firstly, it is unable to explain the M-I phase transition with respect to the LAO thickness. Secondly, the PC sheet carrier density is much larger than the measured one $n_s^{exp} \sim 10^{13}\,e/cm^{-2}$ of the high quality interfaces growth under high oxygen pressure $p_{O_2} \sim 10^{-4}$ mbar. Moreover, it can not describe the microscopic properties of the 2-DEG like the distribution and confinement of the carrier within the interface region.
Numerous ab *initio* calculations were introduced to explain for the I-M phase transition of the defect-free n-type (001) LAO/STO hetero-structure with respect to the LAO thickness, which brings a microscopic description of the 2-DEG and produces more realistic sheet carrier [10, 11]. Despite of that, the sheet carrier density was poorly reproduced, i.e. the theoretical ones are larger than the experimental. The proposed explanations was either electron-phonon interaction or Anderson localizations for the 2D systems [12]. In this study, we investigated the formation of 2-DEG in the hetero-structure when the thickness of the LAO slab was increased in the framework of density functional theory (DFT). On the base of our results, we suggest that the large polar lattice distortion of the LAO partially reduces the intrinsic polar field of the asymmetry LAO slab and hence significantly reduces the carrier density. Besides, the 2-DEG formation is also sensitive to the STO band gap. Band gap underestimation, as a frequent failure of DFT, will be also discussed below as the another reason for the carrier density overestimation.

2. Modeling and Caculation Details

There are two types of n-type (001) LAO/STO hetero-structure models frequently used in previous ab *initio* calculations. The first one is symmetric super-lattice which is composed of a number of LAO u.cs stacked over a number of STO u.cs along (001) direction within the periodic boundary condition. In this model, both LAO and STO are non-stoichiometric, i.e. an extra LaO and TiO_2 layer is respectively added to LAO and STO slab to create two identical n-type interfaces (Fig. 2). This symmetry and non-stoichiometry annul the polarity of the LAO slab and chemically dope a half of electron to each of two interfaces. Consequently, no I-M transition could be observed on the symmetric model and the 2-DEG is controlled by the chemistry of the model rather than the polarity of LAO. The second is the one with stoichiometric and asymmetric LAO slab and a vacuum region added to break the periodicity along the (001) (Fig. 2). This type of model allows us to investigate the LAO polarization effect and an intimate to the real hetero-structure films.

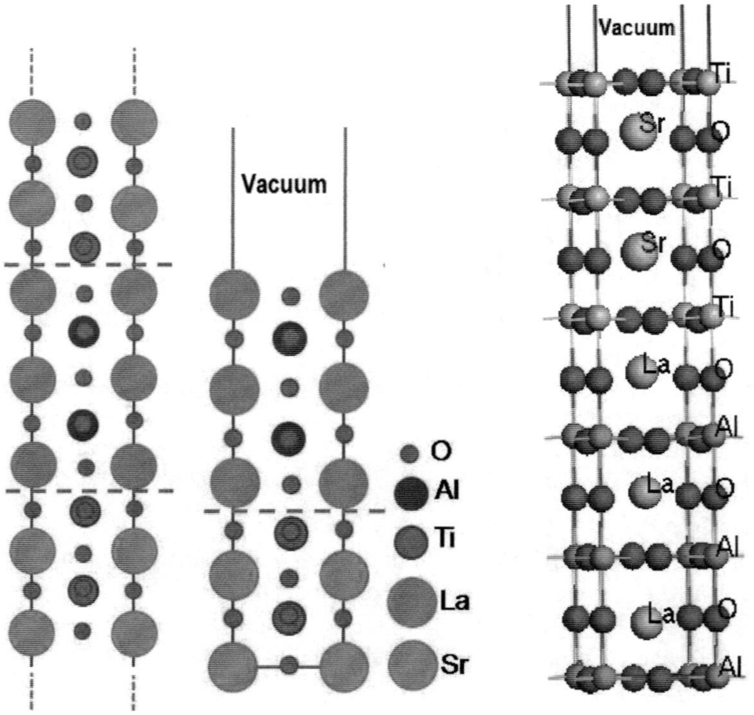

Figure 2. (a) The non-stoichiometric n-type LAO/STO supperlattice; (b) The stoichiometric n-type LAO/STO where the LAO and the STO are separated by a thick vacuum (The dash line denotes the LaO/TiO_2 interface); (c) The LAO/STO modeled slab including 3-ML LAO deposited on 2.5-ML STO substrate.

In order to study the formation of the 2-DEG versus the LAO thickness, we utilized the second model. Each couple of LaO and AlO_2 layers in LAO slab and the SrO and TiO_2 layer in STO slab along the (001) direction is called a monolayer (ML). The STO slab in our model is composed of 2 MLs and an extra TiO_2 atomic layer (see Fig. 2). It is noted that unlike LAO, due to the neutrality of (001) atomic layers in the STO, the non-stoichiometry of the STO slab gives no extra charge. The thickness of the LAO slab increases from 1 to 4 MLs.
Within the framework of DFT, the exchange-correlation term in the ground state equations of the studied systems were treated by the generalized gradient approximation (GGA), i.e. PBE functional [13]. The plane wave basis set was chosen with a k-mesh of 24x24x200 corresponding to the 4ML slab and the energy cut-off was $E_{\text{cutoff}} = 340$ eV. The electron Coulomb potential was an ultra-soft pseudo-potential. The optimized structure (relaxed structure) was the one of the minimum on the potential surface of the system where the first derivative of energy along any direction (force) is zero. In order to find such point, we utilized the BFGS optimization scheme [14] with the convergence criteria of the max force ~ 0.05 eV/Å, of the max displacement ~ 0.005 $\mathring{A}$ and the starting point is the bulk structure of the LAO (unrelaxed structure).

3. Results and Discussions

Firstly, it is useful to simply discuss about the formation of 2DEG at the LAO/STO (001) hetero-interface from the electrostatic view point. Each added LAO ML corresponds to the addition of a polar electric field,

$$E_{\text{polar}} = \frac{2\pi e}{\varepsilon_{\text{LAO}} a_{\text{STO}}^2}, \tag{1}$$

which increases the electrostatic potential by an amount of $\Delta = \frac{E_{\text{polar}}}{a_{\text{LAO}}}$. Using $a_{\text{STO}} = 3.905\mathring{A}$ and $a_{\text{LAO}} = 3.789\mathring{A}$, we have: $E_{\text{polar}} = \frac{5.94V/\mathring{A}}{\varepsilon_{\text{LAO}}}$ and $\Delta V = \frac{22.5\text{eV}}{\varepsilon_{\text{LAO}}}$. Here ε_{LAO} is the dielectric constant of the LAO which evaluates the polarization ability of the LAO to avoid the applied electric field. According to the PC model, without the LAO reaction, the accumulation of the polar electric field would raise up the potential by an amount of 22.5 eV per ML. The increased potential, in turn, raises the valence band (VB) top up an amount larger than the STO band gap ($E_g^{\text{STO}} = 3.2$ eV) so that the LAO valence band will overlap the STO conduction band (CB) (see Fig. 3).

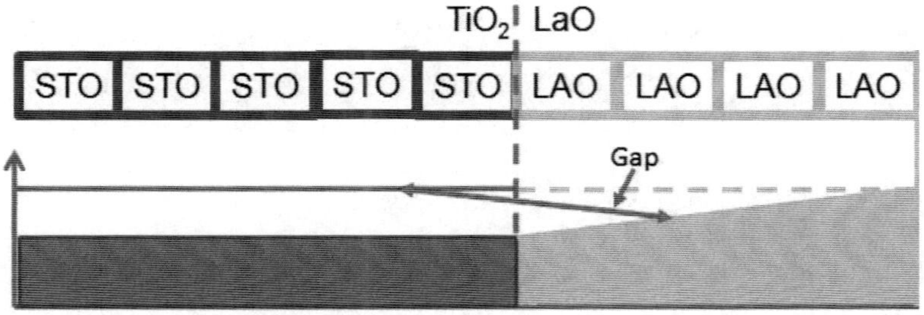

Figure 3. The scheme of the LAO/ STO valence band (VB) and conduction band (CB). The shaded blue part shows the STO VB and the shaded yellow part shows the LAO VB. The blue line presents the bottom of the STO CB.

Consequently, charge transfer from the LAO to the STO, and a 2-DEG will occur when a monolayer of the LAO is added onto a bared STO substrate. In fact, the LAO does strongly respond to the polar field via electronic and ionic polarization. If we use the experimental bulk dielectric constant $\varepsilon_{LAO} = 26$, the potential change is ~ 0.87 eV, indicating that more than 3 MLs of LAO might be required to switch on the 2-DEG conductivity. This estimation agrees quite well with the experimental observations of the MI phase transition at the LAO critical thickness $\sim$ 4 u.cs [2].

Now we turn to discuss our first principle results. To clarify the role of lattice distortion, the electronic structures of both relaxed and unrelaxed hetero-structures were calculated. We observed metal phase for unrelaxed hetero-structure with only 1 LAO ML which indicates that electronic polarization is not enough to inhibit the polar field. On the contrary, insulator phase sustains in the relaxed hetero-structures with the LAO thickness up to 3 MLs. Fig. 3 and 3 show the electrostatic potential oscillation along the (001) direction within the unrelaxed and relaxed 3-ML LAO hetero-structures. The average potential (the straight line) increases 0.5 eV/ML in the case of the unrelaxed structure while it increases more slightly in the case of the relaxed structure ($\sim$0.2 eV/ML). It is suggested that the large polar lattice distortion strongly reduce the potential divergence, hence, screening the charge transfer and delaying the metal phase. The polar field with lattice distortions can be roughly estimated from the ionic charges Z_i and positions of the ith atom z_i (positions along the direction perpendicular to the interface) using $D = \sum_i Z_i z_i$ [15]. For the 3 LAO MLs, $D = 2.65$ e$\mathring{A}$ which is about 45% of the unrelaxed polar field. That means the polar lattice distortion reduces the polar field by 55%.

Our calculation results are in good agreement with the experiment to obtain the switch-on of the metal phase at the LAO thickness of 4 MLs. Fig. 3 and 3 show the calculated band structure

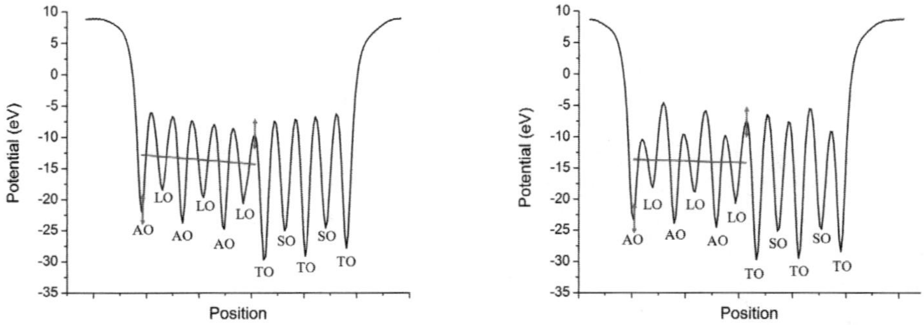

Figure 4. Electrostatic potential oscillations versus layer positions in (a) an unrelaxed and (b) a relaxed structure of 3-ML LAO hetero-structure. LO, AO, TO and SO denote the LaO, AlO_2, TiO_2 and SrO layers, respectively. The red straight line presents the average potential in the LAO.

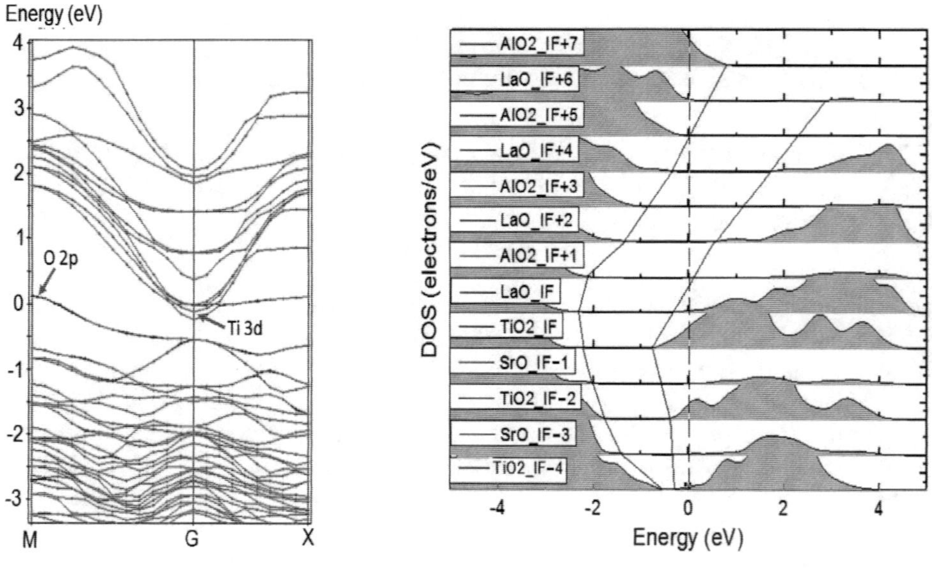

Figure 5. (a) Band structure and (b) layer-projected density of states of the 4-ML LAO hetero-structure.

and layer-projected density of states (DOS) of the 4-ML LAO hetero-structure, respectively. Three sub-bands at the bottom of the CB overlapping the Fermi level originate from the Ti $3d_{xy}$ orbital of the interface (IF) TiO_2 layer (the lowest band), the sub-interface (IF-2) TiO_2 layer and the surface (IF-4) TiO_2 layer (the highest band). The only one sub-band at the top of the VB overlapping the Fermi level is composed of the O 2p-like states locating on the capping AlO_2 layer (IF+7). This indicates that the electrostatic field of 4-ML LAO is large enough to saturate the polarization of the LAO slab and break down the dielectric behavior. The electrons are transferred from the AlO_2 capping layer to the STO slab, becoming free carriers.
The distribution of conducting electrons over atomic layers of the STO slab in the 4-ML hetero-

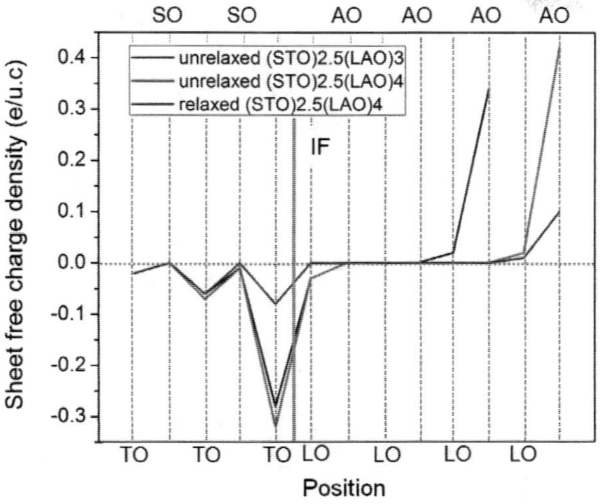

Figure 6. Free charge density versus layer positions in the 4-ML hetero-structure. IF denotes the TiO_2/LaO interface.

structure is shown in Fig. 6. The sheet carrier density quickly reduces from the IF, i.e. 50% of the carrier on the IF layer. This exhibits the two dimensionality of the electron gas at the hetero-interface which was experimentally observed to be confined within several nanometers from the interface. The total sheet carrier density is $n_s = 0.45$e/u.c $\sim 2.97 \times 10^{14}$e/cm^2 for the unrelaxed 4ML hetero-structure. The lattice distortion partially screens the charge transfer, reducing the sheet carrier by 4 times, i.e. $n_s = 0.11$ e/u.c $\sim 7.26 \times 10^{13}$ e/cm^2. However, this value is still several times larger than the experimental ones. In order to explain for this, the failure of local density approximation (LDA) in estimating the STO band gap might be taken into account. As shown above, in order to breakdown the dielectric, it is necessary for the polar field to be accumulated enough for the top of the LAO VB to overlap the bottom of the STO CB. The more the VB overlaps the CB, the more free carriers are transferred and the denser the 2-DEG becomes. According to this, the band gap underestimation, as a frequent failure of the LDA and the semi-LDA methods, gives rise to the overestimation of the CB-VB overlapping and hence, overestimates the carrier density.

4. Conclusion

To conclude, we observed the formation of 2-DEG at the n-type (001) LAO/STO hetero-structure on the base of the DFT calculation results. According to this, the strong polar lattice distortion and electronic polarization can resist the intrinsic polar field of the (001) asymmetry LAO slab of the thickness up to 3 MLs. The dielectric is broken down by the polar field of 4ML slab, which well agrees with experiments. The strong lattice distortion delays the switch-on of the 2D conductivity and reduces the sheet carrier by 4 times in comparison with the PC one. Our *ab initio* calculations still overestimate the sheet carrier density due to the band gap underestimation failure of the semi-LDA methods.

Journal of Physics: Conference Series **537** (2014) 012007 doi:10.1088/1742-6596/537/1/012007

Acknowledgment
This research is funded by Vietnam National Foundation Foundation for Science and Technology Development (NAFOSTED) under grant number **103.02-2012.73**.

References

[1] A. Ohtomo and H. Y. Hwang, Nature **427**, 423 (2004).
[2] S. Thiel, G. Hammerl, A. Schmehl, C. W. Schneider, J. Mannhart, *Science* **313**, (2006) 1942.
[3] A. Brinkman, M. Huijben, M. Van Zalk, J. Huijben, U. Zeitler, J. C. Maan, W. G. Van Der Wiel, G. Rijnders, D. H. A. Blank and H. HIlgenkamp, *Nature Materials* **6**, (2007) 493.
[4] N. Reyren, S. Thiel, A. D. Caviglia, L. Fitting Kourkoutis, G. Hammerl, C. Richter, C. W. Schneider, T. Kopp, A. S. Rüetschi, D. Jaccard, M. Gabay, D. A. Muller, J. M. Triscone, J. Mannhart, *Science* **317**, (2007) 1196.
[5] Ariando, X. Wang, G. Baskaran, Z. Q. Liu, J. Huijben, J. B. Yi, A. Annadi, A. Roy Barman, A. Rusydi, S. Dhar, Y. P. Fend, J. Ding, H. Hilgenkamp and T. Venkatesan, *Nat. Commun.* **2**, (2011) 188.
[6] Lu Li, C. Richter, J. Mannhart and R. C. Ashoori, *Nat. Phys.* **7**, (2011) 762.
[7] M. Sing, G. Bermer, K. Goβ, A. Müller, A. Ruff, A. Wetscherek, S. Thiel, J. Mannhart, S. A. Pauli, C. W. Schneider, P. R. Willmott, M. Gorgoi, F. Schafers, and R. Claessen, *Phys. Rev. Lett.* **102**, (2009) 176805.
[8] Naoyuki Nakagawa, Harold Y. Hwang and David A. Muller, *Nature Materials* **5**, (2006) 204.
[9] P. R. Wilmott, S. A. Pauli, R. Herger, C. M. Schlepütz, D. Martoccia, B. D. Patterson, B. Delley, R. Clarke, D. Kumah, C. Cionca, and Y. Yacoby, *Phys. Rev. Lett.* **99**, (2007) 155502.
[10] Hanghui Chen, Alexie Kolpak, and Sohrab Ismail-Beigi, *Phys. Rev. B* **82**, (2010) 085430.
[11] Massimiliano Stengel, *Phys. Rev. Lett.* **106**, (2011) 136803.
[12] Z. S. Popović, S. Sapathy, and R. M. Martin, *Phys. Rev. Lett.* **101**, (2008) 256801.
[13] J. P. Perdew, K. Burke, and M. Ernzerhof, *Phys. Rev. Lett.* **78**, (1997) 1396.
[14] B. G. Pfrommer, M. Cote, S. G. Louie, M. L. Cohen, *J. Comput. Phys.* **131**, (1997) 233.
[15] Rossitza Pentcheva and Warren E. Pickett, *Phys. Rev. Lett.* **102**, (2009) 107602.

Exact Solution of Random Graphs for Cluster Fragmentation

D T Nga[1], N T Phuong Lan[2], D C Nghia[2] and P Désesquelles[3]

[1]Institute of Physics, Vietnam Academy of Science and Technology, 10 Dao Tan, Thu Le, Ba Dinh, Ha Noi, Vietnam

[2]Hanoi Pedagogical University 2, Nguyen Van Linh Street, Vinh Phuc Province, Vietnam

[3]University Paris-Sud and CSNSM CNRS/IN2P3, 15 rue G. Clemenceau, 91405 Orsay, France

E-mail: `lanntphuong@gmail.com`

Abstract. We present the exact solution of a combinatorial fragmentation model and we show how it can be used as a touchstone for the fragmentation of atomic clusters. This model, random graphs (RG), also called mean field percolation, displays a phase transition. In this model, the clusters are solely described as connected entities called nodes. The connections, called bonds, can be active of broken. We have established the algebraic formulas of the probability of all the fragmentation channels. The results depend on the number of nodes and of the number of broken bonds. Using RG, we show example where information was deduced from fragmentation of systems consisting of finite sets of nodes.

1. Introduction

Random Graphs (RG) were originally introduced in 1959-1960 by Erdös and Rényi [1, 2]. The RG are the most general mathematical models for compound systems of sets of objects in interaction. These models consist only of nodes connected by bonds. At first, this model was used to prove deterministic properties of graphs. Later, interest in random graphs of a different nature arose. Due to the increase of computer power, it has become possible to study real-life networks. Recent applications of RG model are shown in the field of problems of network robustness and of epidemics spreading on contact networks. Typical examples are the termite nest chambers [3], electricity networks or the distribution of diseases [4]. For the physical systems, the nodes represent nucleons, atoms and molecules. The physical information was deduced, using RG, for nuclear spinodal decomposition [5, 6], nuclear fragmentation [7, 8], cluster fragmentation [9, 10] and for the fragmentation of the carbon nucleus into three alpha particles, which could sign the Hoyle state [11].

In this paper, the RG are used as a rigorous formalism allowing to describe and to interpret the fragmentation of clusters. Furthermore, its allow to distinguish the physical correlations, providing information about the system, from the trivial correlations, due to the combinatorial constraints or to conservation laws.

2. Algebraic Formulas of Random Graphs

We consider a homogeneous cluster of S undistinguishable nodes. Each bond between a pair of nodes is associated to the same energy. We consider that the breaking of a bond requires a unit

Journal of Physics: Conference Series **537** (2014) 012008 doi:10.1088/1742-6596/537/1/012008

excitation energy. A cluster is said to be in its ground state when every node pairs is connected by a bond, hence, the ground state energy of S-node clusters is:

$$E_{\mathrm{g}}(S) = -\frac{S(S-1)}{2} \quad (1)$$

A cluster is in an excited state if the number of broken bonds is not enough to break it. In excited states, all nodes are still connected directly or indirectly via the other nodes. When enough bonds are broken, the cluster separates into a set of fragments called a partition. In the example shown in Fig. 1, the parent cluster has 7 nodes (dots) and 21 bonds (lines). After the breaking of 15 bonds (dashed and dotted lines), the parent cluster decays into a partition with 2 fragments (black dots and open dots).

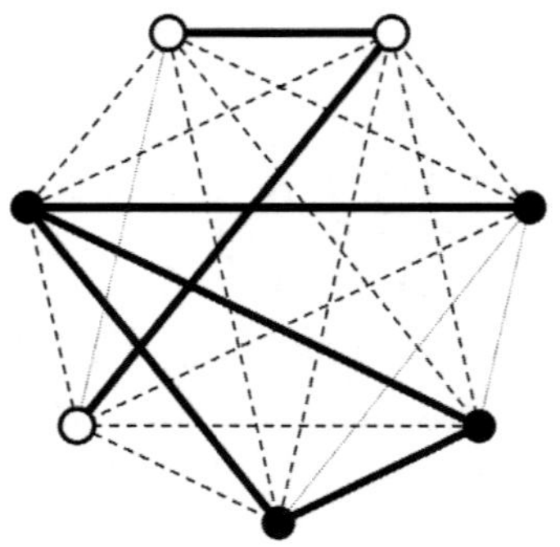

Figure 1. Example of RG fragmentation. The dots represent the nodes and the full lines represent the unbroken bonds. The excitation energy ($E^* = 15$) is shared between the binding energy of the partition ($E_{\mathrm{b}} = 12$, dashed lines) and the individual excitation energies of the fragments ($E_1^* = 2$, for the black dot fragment and $E_1^* = 1$ for the open dot fragment).

A fragment partition [5] will be represented as a vector $\mathbf{n} = (n_1, ..., n_s, ..., n_S)$, whose component n_s indicates the number of fragments with size s. The sum of components $M = \sum_s n_s$, is called the multiplicity and the mass (size of the system) conservation reads $\sum_s s\, n_s = S$. The binding energy $E_{\mathrm{b}}(\mathbf{n})$ of a given partition is the minimal energy to create it. In this model, it is the difference between the number of bonds of the parent cluster and of the sub-clusters in their ground states. Using eq.(1), we can show that it can be also written:

$$E_{\mathrm{b}}(\mathbf{n}) = \frac{1}{2}(S^2 - \sum_{s=1}^{S} s^2\, n_s) \quad (2)$$

When the initial excitation energy E^* (number of broken bonds) injected into the parent cluster is larger than the binding energy of a given partition $\mathbf{n}$, the remaining energy is distributed among the fragments as individual excitation energy

$$E^* = E_{\mathrm{b}}(\mathbf{n}) + \sum_{i=1}^{M} E_i^* \quad (3)$$

where E_i^* is the excitation energy of fragment i. For a given initial excitation energy E^*, the microcanonical weight of a fragmentation partition depends on two factors which respectively

represent the number of ways to arrange the nodes among the fragments and on the number of ways to distribute the excitation energy among the fragments.
The first factor is the combinatorial factor representing the number of ways to allocate S nodes to the fragments. There are $S!$ number of ordering of all the nodes. However, the permutation of nodes inside a fragment does not change the partition nor does the permutation of equal size fragments. Thus this factor can be written as:

$$w_{\text{comb}}(\mathbf{n}) = \frac{S!}{\prod_{s=1}^{S} s!^{n_s}\, n_s!} \tag{4}$$

The second factor is related to the number of the ways to distribute the remaining energy among the fragments of the partition. There are different ways to sort the excitation energy among the fragments. A distribution of the total excitation energy is characterized by a vector $\mathbf{E}$ with M dimensions which components are the E_i^* with the constraints:

$$E_i^* \in [0, (s_i - 1) \times (s_i - 2)/2] \qquad \text{and} \qquad \sum_i E_i^* = E^* - E_{\text{b}}(\mathbf{n})$$

Moreover, for a given fragment i, the density of states $\rho(s_i, E_i^*)$ is equal to the number of ways to choose E_i^* broken bonds among $s_i \times (s_i - 2)/2$ bonds without breaking the fragment. Hence, the product of the fragments level densities has to be taken into account in the evaluation of the partition weight. Finally, the partition weight corresponding to the given E^* can be expressed as:

$$w(\mathbf{n}, E^*) = w_{\text{comb}}(\mathbf{n}) \sum_{\mathbf{E}} \prod_{i=1}^{M} \rho(s_i, E_i^*) \tag{5}$$

All the observables and the thermodynamic features of the model can be derived from this equation. Especially, equation (5) allows to determine the partition probabilities as a function of the initial excitation energy. It is calculated by:

$$P(\mathbf{n}|E^*) = \frac{w(\mathbf{n}, E^*)}{\sum_{\mathbf{n}} w(\mathbf{n}, E^*)} \tag{6}$$

where the sum is over all possible partitions. The density of states $\rho(S, E^*)$=$w((0, ..., 0, 1), E^*)$ can easily be computed noticing that it is the total number of ways to choose E^* broken bonds among $-E_{\text{g}}(S)$ bonds minus the sum of all the partition weights. Thus we have:

$$\rho(S, E^*) = \binom{-E_{\text{g}}}{E^*} - \sum_{\substack{\mathbf{n} \\ \mathbf{n} \neq (0,...,0,1)}} w_{\text{comb}}(\mathbf{n}) \sum_{\mathbf{E}} \prod_{i=1}^{M} \rho(s_i, E_i^*) \tag{7}$$

Here s_i and E_i^* are smaller than S and E^* thus the equation can be used to calculate recursively the densities of states.
Real life systems may also be composed of different types of "nodes" (for example protons and neutrons in the nucleus or atomic species in molecules). For T types, the combinatory factor of the heterogeneous RG model is given by [12]:

$$w_{\text{comb}}(\mathbf{N}) = \frac{\prod_{t=1}^{T} S_t!}{[\prod_{s=1}^{\max(S_1,...,S_T)} s!^{\sum_t n_{ts}}][\prod_{s_1=0}^{S_1} \cdots \prod_{s_T=0}^{S_T} N_{s_1...s_T}!]}. \tag{8}$$

where S_t the number of type t nodes, n_{ts} the number of fragments containing s nodes of type labeled t and $N_{s_1...s_T}$ the number of fragments with s_1 nodes of type 1 , . . . , s_T nodes of type T.

3. Results

3.1. Diagram for fragmentation channel probabilities

We first present the results obtained from the exact equations of RG partition probabilities as a function of the excitation energy (number of broken bonds). Figures 2 and 3 show the diagram for fragmentation channel probabilities of C_5 ($S = 5$) and C_4H ($S_1 = 4$ and $S_2 = 1$) clusters, respectively. The fragmentation of the same atomic clusters has been studied experimentally by [13] and [14]. The highly excited C_5 and C_4H clusters may decay into seven and twelve fragmentation channels (partitions), respectively. These figures show the thresholds of appearance of the fragmentation channels as well as the dominant partition corresponding to a domain of excitation energy. The C_5 and C_4H clusters do not dissociate up to $E^* = 3$. We note that the partitions having the same number of fragments cover approximately the same range of excitation energy. For example, the two fragment channel: C_3/C_2 and C_4/C of C_5 cluster; C_4/H, C_3H/C, C_3/CH and C_2H/C_2 of C_4H cluster, appear in the range of excitation energy from 4 to 8. In the energy region from 7-9, only the fragmentation channels leading to three fragments play a significant role.

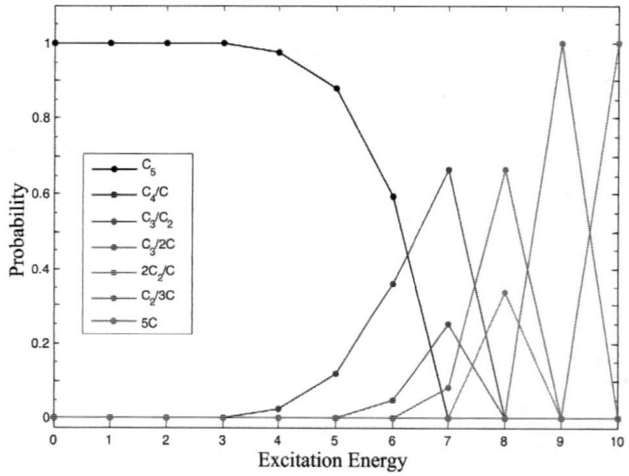

Figure 2. Partition probabilities as a function of the number of broken bonds for the fragmentation of C_5 from RG model.

3.2. Multiplicity probabilities as a function of the excitation energy

The objective of this section is to present the matrix $P(M|E^*)$ representing multiplicity probabilities as a function of the excitation energy E^* to show the correlation between M and E^*. The probability $P(M|E^*)$ is the sum of the equal multiplicity partition probabilities. These probabilities satisfy the following normalization:

$$\forall E^*, \qquad \sum_{M=1}^{S} P(M|E^*) = 1 \tag{9}$$

For a fragmentation partition of the system with size S leading to M fragments, the total minimal number of bonds necessary to connect all the nodes in the fragments is $L_{\text{min}} = S - M$ (each fragment is a linear chain). The maximal excitation energy to produce this fragmentation is thus:

$$E^* = -E_{\text{g}} - (S - M) \tag{10}$$

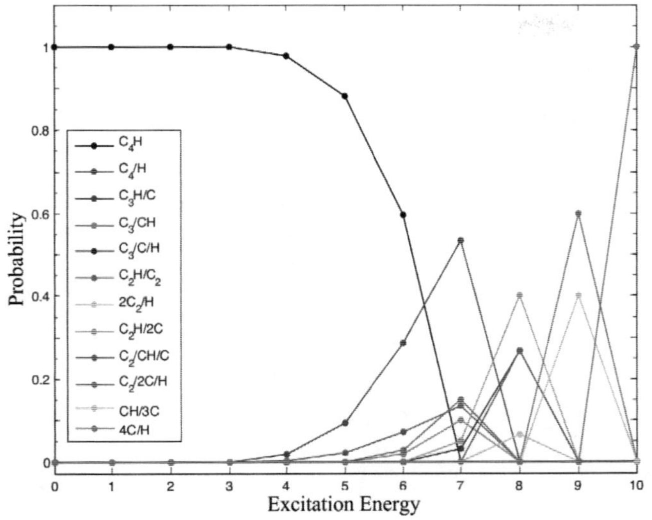

Figure 3. Partition probabilities as a function of the number of broken bonds for the fragmentation of C_4H from RG model.

Using equations (1) and (10), we deduce the relation between M and E^* for the RG model:

$$M = E^* - \frac{S(S-3)}{2} \tag{11}$$

Equation (11) shows that the correlation between M and the maximal excitation energy is linear with a threshold energy $E^*_{\text{threshold}} = S(S-3)/2$.
Figures 4 and 5 represent $P(M|E^*)$ for the system $S = 5$ which interests us and for a larger system $S = 15$ on which the effect is more visible. As can be seen, at high multiplicity, the dispersion according to the excitation energy is very small, the correlation thus becomes linear (see figure 5). The same behavior has been evidenced for atomic clusters by Chabot et al..

3.3. Convergence of a RG Monte Carlo simulation

We have studied the convergence of the Monte Carlo version of the RG model towards the exact algebraic solution. We illustrate here in the case of the system with the size $S = 7$ nodes in the ground state (E_g = -21) with the excitation energy $E^* = 15$. The probabilities of the 8 possible partitions are calculated for 3 values of the total number of Monte Carlo events. For each event, the Carlo Monte program breaks randomly 15 bonds among the 21 bonds. The corresponding partition counting rate is indented and the program starts again. The probability of a partition obtained by Monte Carlo simulation, is equal to its counting rate divided by the total number of events. In Fig. 6 are represented the results for 10^2, 10^3 and 10^4 events. We see that when the total number of events increases, the probabilities obtained by the Monte Carlo simulation converge rapidly towards the probabilities calculated algebraically.

4. Conclusions

We have applied the Random Graphs model to describe the fragmentation of excited homogeneous and heterogeneous systems. We have also established the algebraic equations of the RG model and used that as a tool to evidence properties or correlations in physical systems submitted to fragmentation.

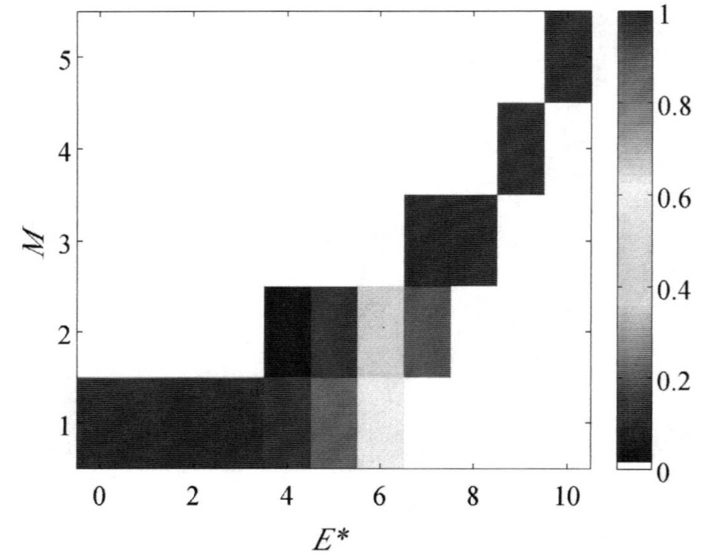

Figure 4. Fragmentation probabilities as a function of the multiplicity M and the excitation energy E^* of the system containing 5 nodes from RG model.

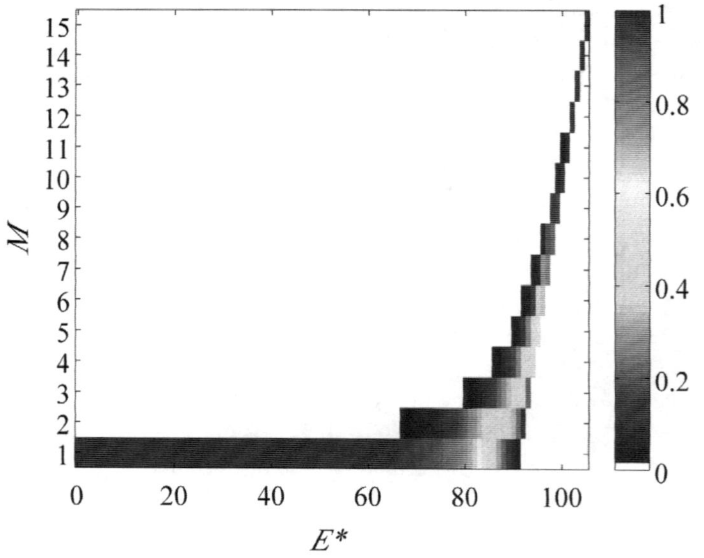

Figure 5. Fragmentation probabilities as a function of the multiplicity M and the excitation energy E^* of the system containing 15 nodes from RG model.

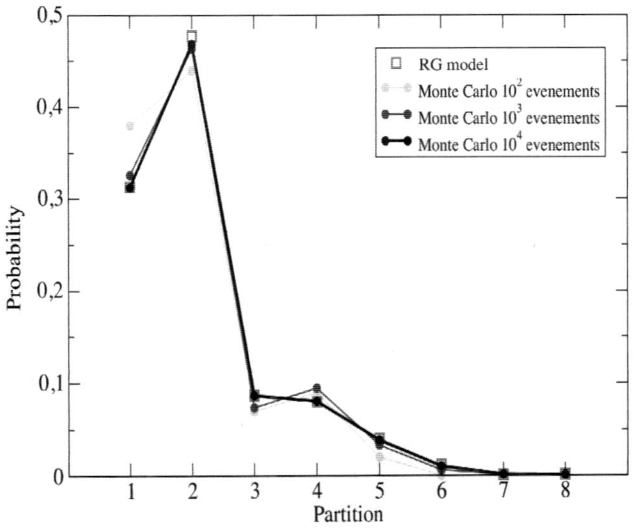

Figure 6. Probabilities of the 8 possible partitions obtained by Monte Carlo simulation and by the algebraic equations of RG model, for the system containing 7 nodes in the ground state with excitation energy $E^* = 15$ (as in Fig. 1).

Acknowledgment

Financial supports of National Foundation for Science and Technology Development (NAFOSTED) of Vietnam through Grant No.103.01-2013.25 and VAST through the project supporting young researchers are gratefully acknowledged.

References

[1] P. Erdös and A. Rényi. On random graphs. I. Publ. Math. Debrecen, 6 (1959) 290-297.
[2] P. Erdös and A. Rényi. On the evolution of random graphs. Bull. Inst. Internat. Statist., 38 (1961) 343-347.
[3] P. Désesquelles, D.-T. Nga, A. Cholet, Journal of Physics: Conference Series, 410 (2013) 012058.
[4] M. Gonzalez, Contact Networks of Mobile Agents and Spreading Dynamics, Logos Verlag, Berlin (2006).
[5] P. Désesquelles, Phys. Rev. C, 65 (2002) 034603.
[6] B. Borderie et al., Nucl. Phys. A, 734 (2004) 495-503.
[7] P. Désesquelles et al., Phys. Rev. C 62 (2000) 024614.
[8] P. Pawlowski et al., Phys. Rev. C 57 (1998) 1771-1777.
[9] S. Diaz-Tendero, International Journal of Mass Spectrometry, 252 (2006) 126-132.
[10] K. Beroff, Nucl. Instr. and Meth. A, 267 (2009) 866.
[11] P. Schuck et al., Progress in Particle and Nuclear Physics, 59 (2007) 285-304.
[12] P. Désesquelles, Physics Letters B, 698 (2011) 284-287.
[13] T. Tuna et al., Journal of Chemical Physics, Vol. 128, (2008) 124312.
[14] M. Chabot et al., Astronomy and Astrophysics, Vol. 524 (2010) A39.

IWTCP1 & NCTP38 IOP Publishing
Journal of Physics: Conference Series **537** (2014) 012009 doi:10.1088/1742-6596/537/1/012009

Study of Thermodynamic Properties of AuCu Superlattice by the Statistical Moment Method

Vu Van Hung[1], Cao Huy Phuong[2] and Nguyen Thi Hoa[3]
[1]Vietnam Education Publishing House, 81 Tran Hung Dao, Hoan Kiem District, Hanoi, Vietnam
[2]Hung Vuong University, Nguyen Tat Thanh Street, Viet Tri City, Phu Tho Province, Vietnam
[3]University of Transport and Communications, Lang Thuong Ward, Dong Da District, Hanoi, Vietnam

E-mail: caohuyphuongvlpt@yahoo.com.vn

Abstract. We have investigated the thermodynamic quantities of AuCu superlattice using the statistical moment method in the statistical physics. The free energy, thermal lattice expansion coefficients, mean-square atomic displacements, and specific heats at the constant volume and those at the constant pressure, C_v and C_p, are derived in closed analytic forms in terms of the power moments of the atomic displacements. The present analytic formulas including the anharmonic effects of the lattice vibration give the predicted values of these quantities.

1. Introduction

The study of the thermodynamic properties of metals, alloys and metallic superlattices are now of great importance for the understanding of structural phase transformations as well as for the technological applications. The composition and temperature range of the AuCu-type superlattice formation in the Au-Cu-Ag ternary system was determinded by X-ray powder diffraction [1].

The ternary system of gold-copper-silver is characterized by an ordering region in which the copper-rich and silver-rich phases co-exist. As shown in Fig 1, the AuCu I superlattice form in which alternate (001) planes contain only copper or only gold atoms and the crystallographic unit cell of the AuCu II is 10 AuCu I cells in length.

In order to understand the temperature dependence of the thermodynamic quantities, it is highly desirable to establish an analytical method which enables us to evaluate the free energy of the system taking into account both the anharmonicity and quantum mechanic effect of the lattice vibration. So far the numerical calculation methods, such as the molecular dynamic and Monte Carlo simulation techniques have been presented. However, it is generally difficult to get simple algebraic formula between the thermodynamic quantities and physical insight of the phenomena, within the non-analytical numerical simulation studies.

The application calculations using the statistical moment method (SMM) have been performed for ordinary cubic metals[2] and for the bcc alloys within the studies of thermodynamic properties[3]. Recently much attention has been paid to alloy systems.

It is the purpose of the present article to study the thermodynamic properties in AuCu II superlattice using the statistical moment method[4, 5].

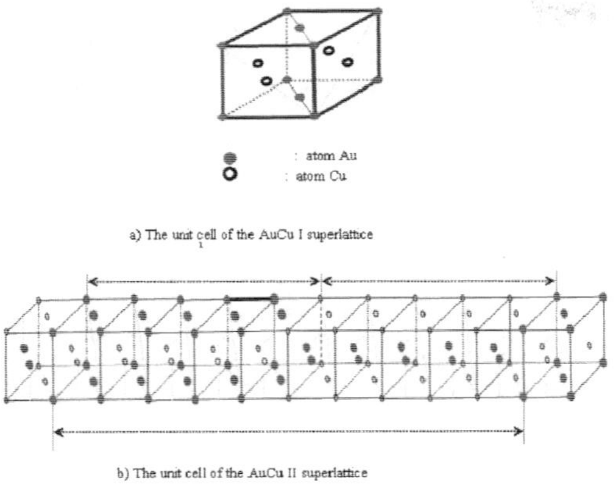

Figure 1. The unit cell of the AuCu I and AuCu II superlattice. Open circles and solid circles represent the gold and copper atoms, respectively.

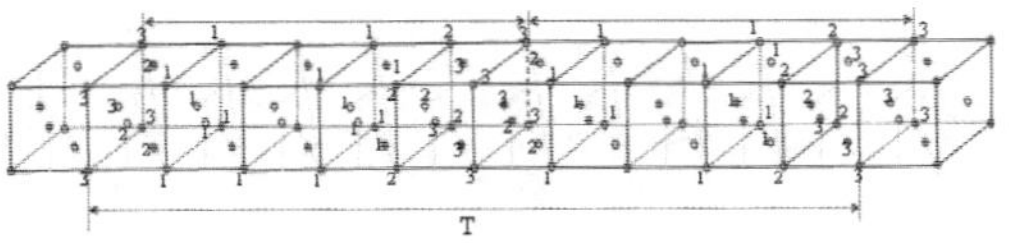

Figure 2. Lattice positions of Au and Cu atoms in the period of the AuCu II superlattice.

We calculate the thermodynamic quantities of AuCu II superlattice taking into account the anharmonicity of thermal lattice vibrations. The free energy, nearest-neighbour distance, thermal expansion, co-efficient and specific heats C_v and C_p of superlattice are derived in the term of the zero pressure and the temperature T. The numerical calculations are perform for superlattice AuCu, using the effective pair potential between atoms. The present paper is organized as follows: In sec.2, we present the principles of calculations for thermodynamic quantities of AuCu II superlattice. The numerical calculation results and the related discussions are given in sec.3.

2. Principle of Calculations

In this section, we outline the calculations of the thermodynamic quantities of AuCu II superlattice at finite temperature with the use of the SMM[4, 5].

We consider the AuCu II superlattice with three distinct sublattices i (i= 1,2,3). The concentrations of Au and Cu atoms in the sublattice i

$$C_Au^1 = C_Cu^1 = \frac{3}{10}; C_Au^2 = C_Cu^2 = C_Au^3 = C_Cu^3 = \frac{1}{10} \tag{1}$$

The free energy of the AuCu II superlattice is then obtained by taking into account the configurational entropy S_c,via the Boltzman relation, and written as:

$$\psi = N\left(C_Au^1\psi^1_{Au} + C^2_{Au}\psi^2_{Au} + C^3_{Au}\psi^3_{Au} + C^1_{Cu}\psi^1_{Cu} + C^2_{Cu}\psi^2_{Cu} + C^3_{Cu}\psi^3_{Cu}\right) \tag{2}$$

Where ψ^i_Y(Y=Au,Cu; i=1, 2, 3) is the Helmholtz free energy of an atom Y (Y=Au, Cu) located at the lattice position i (i=1, 2, 3). The general expression of the Helmholtz free energy is given from the SMM[4] as

$$\psi^i_Y = u^i_{oY} + 3\theta x^i_Y + ln(1 - e^{-2x}) \tag{3}$$

Journal of Physics: Conference Series **537** (2014) 012009 doi:10.1088/1742-6596/537/1/012009

where $x_{Y^i} = \frac{\hbar\omega^{iY}}{2\theta}$; $\omega^{iY} = \sqrt{\frac{k^i_Y}{m}}$ and $\theta = k_B T$ Here k_B and m represent the Boltzman constant and atomic mass, respectively. The first term u^i_{oY} denotes the internal energy of the system, and the second-order vibrational constant k^i_YkiY are given by

$$u^i_{oY} = \frac{1}{2}\sum_j \varphi^i_{jY}(|r_j|)k^i_Y = \frac{1}{2}\sum_j \left(\frac{\partial^2 \varphi^i_{jY}}{\partial u^2_{jx}}\right)_{eq} \tag{4}$$

where r_j is the equilibrium position of the jth particle, φ^i_{jY} the interaction potential energy between j^{th} particles and Y particle located at the lattice site i and u_{jx} the atomic displacement of jth particle due to thermal lattice vibrations, and eq denotes the equilibrium interatomic distance. Applying the Gibbs-Helmholt relation and using 2 we find the expression for the energy of the AuCu II superlattice

$$E = N\left[\sum_{i,Y} C^{Y^i} E^i_Y\right] \tag{5}$$

Where $E^i_Y = \psi^i_Y - \theta\left(\frac{\partial \psi^i_Y}{\partial \theta}\right)$ So the specific heat at constant volume C_v has the form

$$C_V = \left(\frac{\partial E}{\partial T}_V\right) \tag{6}$$

$$= \sum_{i,Y} C^i_Y C^{Yi}_V \tag{7}$$

Where $C^{Yi}_V = \frac{\partial E}{\partial T}$
Since the thermodynamic quantities can be expressed in terms of temperature and pressure derivatives of the Gibbs (or Helmholtz) free energy, it can be seen, that the contributions of the free energy of Au and Cu atoms located at the different lattice sites to the thermodynamic properties of the AuCu II superlattice are additive. Therefore, in an analogous way as for finding 5 and 6, the expression of the average displacement of the particle from equilibrium position of AuCu II superlattice can be written in the form

$$y = \sum_{i,Y} C^i_Y y^i_y \tag{8}$$

where y^i_Y is the displacement of the Y atom (Y=Au, Cu) from equilibrium position located at the lattice site i (i=1,2,3). The displacement y^i_Y of AuCu II superlattice can be calculated using the statistical moment method[4, 5]

$$y^i_Y = \sqrt{\frac{2\gamma_Y{}^i\theta}{3\left(k^i_Y\right)^3}A} \tag{9}$$

where fourth order vibrational constant γ^i_Y is defined by:

$$\gamma^i_Y = \frac{1}{12}\sum_j\left[\left(\frac{\partial^4\varphi^i_{jx}}{\partial u^4_{j\alpha}}\right)_{eq} + 6\left(\frac{\partial^4\varphi^i_{jx}}{\partial u^2_{j\alpha}\partial u^2_{j\beta}}\right)_{eq}\right] \tag{10}$$

With $\alpha\sharp\beta = x, y, z$ When the displacement y is determined, we can find the nearest neighbor distance:

$$a = a_o + y^i_Y \tag{11}$$

and the linear thermal expansion coefficient of AuCu II superlattice

$$\alpha_T = \frac{k_B}{a_o}\frac{da}{d\theta} = \frac{k_B}{a_o}\sum_{j,Y} C_Y^i \left(\frac{dy_Y^i}{d\theta}\right) \tag{12}$$

where a_o is the nearest neighbor distance of AuCu II superlattice at the temperature $T = 0$.

3. Results and Discussion

In order to calculate the thermodynamic quantities of AuCu II superlattice, we will use the m-n type of Lennard- Jones potertial between two atoms of a metal[6]

$$\varphi(r) = \frac{D}{n-m}\left[m\left(\frac{r_o}{r}\right)^n - n\left(\frac{r_o}{r}\right)^m\right] \tag{13}$$

Where D, r_o are determined from the experimental data and m, n are determined by the empirical way. For the metal Au, Cu, the potential parameters D, r_o, m and n are taken from [5] and are presented in 3. These parameters are determined so as to fit the experimental lattice constants and cohesive properties.

Metals	m	n	$r_o\,(A^o)$	$D/k_B\,(A)$
Au	5.5	10.5	2.8751	4683.0
Cu	5.0	8.0	2.5487	3401.1

Using these effective potentials, one can find the nearest neighbor distance at temperature T=0K. Then, we calculate the vibration constant at the temperature T=0K In 3, we present the nearest neighbor distances of the pure metals Au, Cu and AuCu II superlattice calculated by the SMM formalism.The nearest neighbor distances of AuCu II superlattice are smaller than those of the pure metal Au and are larger than those of the pure metal Cu.

T(K)	100	200	300	400	500	600	700	800
$a(Ao)$:								
-AuCu	2.7483	2.7535	2.7588	2.7642	2.7697	2.7753	2.7810	2.7869
-Au	2.8458	2.8498	2.8539	2.8581	2.8623	2.8666	2.8710	2.8755
-Au[7]								
$\alpha_T(10^{-5K^{-1}}$								
-AuCu	1.85	1.91	1.94	1.98	2.02	2.06	2.11	2.16
-Au	1.39	1.43	1.48	1.53	1.58	1.4	1.70	1.76
-Au[7]	1.15	1.34	1.41	1.45	1.50	1.54	1.59	1.5
$C_p(\frac{Cal}{mol.K})$								
-AuCu	4.99	5.80	6.02	6.13	6.20	6.27	6.33	6.38
-Au	5.12	5.81	6.00	6.11	6.20	6.31	6.40	6.48
-Au[7]	5.12	5.84	6.07	6.18	6.28	6.40	6.52	6.65

In 4, we present the temperature dependence of the heats capacity at constant volume C_V of AuCu II superlattice and Au pure metal, together with the experimental results [7]. Overall good agreements between the calculation and experimental results of Au pure metal are obtained for a wide temperature range. The heats capacity at constant volume C_V of AuCuII superlattice is smaller those of the Au pure metal. We also present in Table 2, the calculated nearest neighbor distances a, linear thermal expansion coefficients α_T and heats capacity at constant pressure C_p of AuCu II superlattice, Au pure metal, together with the experimental results[7, 8]. The thermodynamic quantities (a, α_T and C_p) are calculated as a function of temperature, and are increasing function of the temperatures, in agreement with the experimental results for the Au pure metal[7].

Journal of Physics: Conference Series **537** (2014) 012009 doi:10.1088/1742-6596/537/1/012009

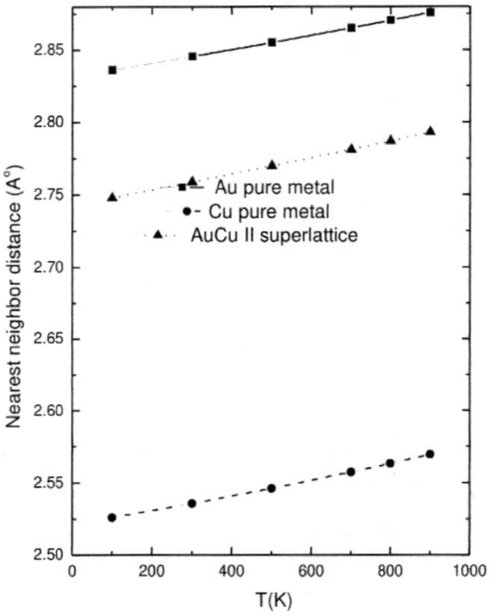

Figure 3. Temperature dependence of the nearest neighbor distance of AuCu II superlattice and Au, Cu pure metals.

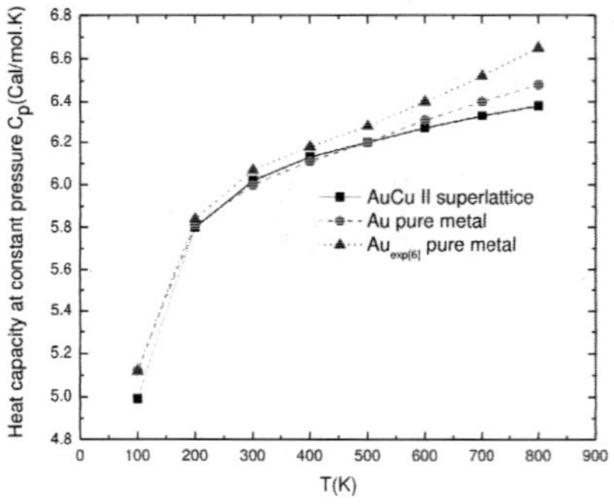

Figure 4. Temperature dependence of the at constant volume C_V of AuCu II superlattice and Au pure metal.

4. Conclusion

Summarizing the above mentioned calculations by the SMM formalism the analytic derivations are quite useful for the numerical calculations of the various thermodynamic quantities. In conclusion, the SMM calculations are performed using the m-n type of Lennard-Jones potential for the AuCu II superlattice with fcc structure. Using the free energy formulas derived by the SMM, we have studied the temperature dependences of the nearest neighbor distance a, thermal expansion coefficient α_T, heats capacities at constant volume and pressure $C_p)C_V)$ of the AuCu II superlattice.

Journal of Physics: Conference Series **537** (2014) 012009 doi:10.1088/1742-6596/537/1/012009

Acknowledgment
This work is supported by the research project No. 103.01.2011.16 of NAFOSTED.

References

[1] Tsutomu U., Yasuo K., and KaTsuhiro Y., J.Dent. Res,*Phys. Rev* **60** (1981) 883.
[2] Vu Van Hung, K. Masuda-Jindo, Pham Dinh TamT,*Journal of Phase Eq.* **22** (2001) 400.
[3] N.Tang and V.V.Hung,*phys. Stat. Sol.(B)* **149** (1988) 511, **161** (1990) 165.
[4] K.Masuda Jindo, V.V.Hung and P.D.Tam,*Phys. Rev.B* **60** (2003) 094301.
[5] V.V.Hung, H.T.Hang and N.T.Hai,*Comm. in phys.* **17** (2007) 64.
[6] Mazomedov M.N., J.Fiz.Khimic,*Phys. Rev* **61** (1987) 1003.
[7] *Americal Institute of physics Handbook* New York, 1963.
[8] David R. Lide, *CRC HandBook of Chemistry and physics* 2005, London- New York - Singapore

Journal of Physics: Conference Series **537** (2014) 012010 doi:10.1088/1742-6596/537/1/012010

Realizations of $SU_q(2)$ Algebra in Terms of q-Deformed Multimode Boson Oscillators

Luu Thi Kim Thanh and Man Van Ngu
Hanoi Pedagogical University No. 2, Xuan Hoa, Phuc Yen, Vinh Phuc, Vietnam
E-mail: luuthikim.thanh888@gmail.com

Abstract. We consider some version of q-deformed multimode boson oscillators. The realization of $SU_q(2)$ algebra in terms of q-deformed multimode boson oscillators which involves q-interference between oscillators of different modes and the realization of $SU_q(2)$ algebra in terms of q- deformed multimode boson oscillators in which each oscillator mode has its own defomation parameter are constracted.

1. Introduction

Quantum groups and quantum algebras have been shown to arise in many problems of current physical and mathematical interest. Much effort is now being devoted to the cosntruction of their representations and recently many realizations have been usefully devised using q-deformations of boson and fermion operators [1, 2, 3].

On the other hand, quantum groups are a subject of great activity at present and although their direct physical interpretation is still lacking, it is of particular importance to study the posible physical implicatios of these deformation. These structures which first emerged in connection with the quantum inverse scattering theory [4], the quantum Yang – Baxter equation .... The algebras may be described as a deformation, depending on one or more parameters of the ordenary Lie algebras [5].

In this paper, we introduction some versions of q-deformed multimode boson oscillators. The realization of $SU_q(2)$ algebra in terms of q-deformed multimode boson oscillators which involves q-interference between oscillators of different modes and the realization of $SU_q(2)$ algebra in terms of q-deformed multimode boson oscillators in which each oscillator mode has its own defomation parameter are constracted.

2. The Realizations of $SU_q(2)$ Algebra in Terms of q-Deformed Multimode Boson Oscillators which Involves q–Interference between Oscillators of Different Modes

In the classical case, the $SU(2)$ generators satisfy the commutation relations

$$\begin{aligned} [J_0, J_-] &= -J_-, \\ [J_0, J_+] &= J_+, \\ [J_+, J_-] &= 2J_0. \end{aligned} \tag{1}$$

Realization of $SU(2)$ algebra in terms of the (ordinary) boson operators as follows

$$\begin{aligned} J_+ &= a_1^\dagger a_2, \\ J_- &= a_2^\dagger a_1, \\ J_0 &= \tfrac{1}{2}(N_1 - N_2). \end{aligned} \tag{2}$$

In this work, we consider a version of deformation which involves some q-interference between oscillators of diferent modes. The creation and annihilation oscillator operators obeying bosonic commutation relations

$$a_i a_j^\dagger - q a_j^\dagger a_i \quad = \delta_{ij} q^N, \tag{3}$$

$$[a_i, a_j] \quad = 0, (i \neq j) \tag{4}$$

where the deformation parameter q being real, and N is the total oscillator number operators,

$$\begin{aligned} N &= \sum_{i=1}^{k} N_i, \\ [N_i, a_j] &= -\delta_{ij} a_j, \\ \left[N_i, a_j^\dagger\right] &= \delta_{ij} a_j^\dagger. \end{aligned} \tag{5}$$

Now the oscillators of different mode enter the theory not quite independently but with some $q$interference, which results from the presence of the factor q^N in the r. h. s. of 3. Equation 3 gives

$$a_i \left(a_i^\dagger\right)^m = q^m \left(a_i^\dagger\right)^m a_i + q^{m-1} m \left(a_i^\dagger\right)^{m-1} q^N. \tag{6}$$

The q-oscillator algebra 3 can be realised in the Fock space spanned by the orthonormalised eigenstates of the oscillator number operators

$$|n_1, n_2, ..., n_k\rangle = \frac{q^{-\frac{n(n-1)}{4}}}{\sqrt{n_1! n_2! ... n_k!}} \left(a_1^\dagger\right)^{n_1} \left(a_2^\dagger\right)^{n_2} ... \left(a_k^\dagger\right)^{n_k} |0\rangle \tag{7}$$

and in this space the following relations hold:

$$\begin{aligned} a_i^\dagger a_i &= q^{N-1} N_i, \\ a_i a_i^\dagger &= q^N (N_i + 1), \end{aligned} \tag{8}$$

and hence

$$\begin{aligned} \sum_{i=1}^{k} a_i^\dagger a_i &= q^{N-1} N, \\ \sum_{i=1}^{k} a_i a_i^\dagger &= q^N (N + k), \end{aligned} \tag{9}$$

where k is the number of modes.

The realisation of Lie algebra in terms of (ordinary) bosons are useful not only as a convenient mathematical tool, but also because of their applications in physics. In the case of quantum algebras it turns out that boson realizations are possible in terms the q-deformed boson operators already introduced above. In the case of $SU_q(2)$ algebra for two mode oscillator eqs. 3 and 4 read:

$$\begin{aligned} a_i a_i^\dagger - q a_i^\dagger a_i &= q^N, \quad (i=1,2) \\ a_1 a_2^\dagger &= q a_2^\dagger a_1, \\ a_1 a_2 &= a_2 a_1. \end{aligned} \tag{10}$$

The realisation of $SU_q(2)$ algebra besed on the q-oscillator algebra 10 can be performed in the Fock space spanned by the orthonormalized of N_i defined as

$$|j,m\rangle = \frac{q^{-j(j-\frac{1}{2})}}{\sqrt{(j+m)!(j-m)!}} \left(a_1^\dagger\right)^{j+m} \left(a_2^\dagger\right)^{j-m} |0\rangle, \tag{11}$$

and the generators can be mapped onto q deformed bosons as follows

$$\begin{aligned} J_+ &= q^{1-N} a_1^\dagger a_2, \\ J_- &= q^{1-N} a_2^\dagger a_1, \\ J_0 &= \tfrac{1}{2}(N_1 - N_2). \end{aligned} \tag{12}$$

In fact, using 10, we can check that these generators satisfy the algebras 1.

3. The Realizations of $SU_q(2)$ Algebra in Terms of q-Deformed Multimode Boson Oscillators in which each Oscillator Mode Has Its Own Deformation Parameter

Now, we consider a version of q-deformed multimode boson oscillators in which each oscillator mode has its own deformation parameter of form:

$$a_i a_j^\dagger - q_i q_j a_j^\dagger a_i = \delta_{ij} q_i^{2N}, \tag{13}$$

$$q_j^{-1} a_i a_j - q_i^{-1} a_i a_j = 0, \tag{14}$$

$$[N, a_i] = -a_i, \tag{15}$$

where N is the total oscillator number operators,

$$\begin{aligned} N &= \sum_{i=1}^{k} N_i, \\ [N_i, a_j] &= -\delta_{\mathrm{ij}} a_j, \\ \left[N_i, a_j^\dagger\right] &= \delta_{\mathrm{ij}} a_j^\dagger. \end{aligned} \tag{16}$$

The basic of the Fock space is defined by repeated action of the creation operators on the vacuum state:

$$\begin{aligned} |n_1, n_2, ..., n_k\rangle &= \frac{1}{\sqrt{n_1! n_2! ... n_k!}} \\ & \prod_{i=1}^{k} q_i^{-n_i \sum\limits_{j>i} n_j - \frac{n_i(n_i-1)}{2}} \left(a_1^\dagger\right)^{n_1} \left(a_2^\dagger\right)^{n_2} ... \left(a_k^\dagger\right)^{n_k} |0\rangle, \end{aligned} \tag{17}$$

where k is the number of the modes. In this space the following relations hold:

$$\begin{aligned} a_i^\dagger a_i &= q_i^{2(N-1)} N_i, \\ a_i a_i^\dagger &= q_i^{2N}(N_i + 1). \end{aligned} \tag{18}$$

In particular, for two mode oscillator equations 13 and 14 read:

$$a_i a_i^\dagger - q_i^2 a_i^\dagger = q_i^{2N}, (i=1,2) \tag{19}$$

$$a_1 a_2^\dagger = q_1 q_2 a_2^\dagger a_1 \tag{20}$$

$$q_2^{-1} a_1 a_2 = q_1^{-1} a_2 a_1. \tag{21}$$

The realisation of $SU_q(2)$ algebra based on the $q$deformed boson oscillator can be performed in the Fock spase spanned by the orthonormalized eigenstates of N_i defined as

$$\begin{aligned} |jm\rangle = & \frac{1}{\sqrt{(j+m)!(j-m)!}} q_1^{-\left[(j+m)(j-m)+\frac{(j+m)(j+m-1)}{2}\right]} q_2^{\frac{-[(j-m)(j-m-1)]}{2}} \\ & \left(a_1^\dagger\right)^{j+m} \left(a_2^\dagger\right)^{j-m} |0\rangle \end{aligned} \tag{22}$$

with the identification

$$\begin{aligned} J_+ &= q_1^{1-N} q_2^{1-N} a_1^\dagger a_2, \\ J_- &= q_1^{1-N} q_2^{1-N} a_2^\dagger a_1, \\ J_0 &= \frac{1}{2}\left[q_1^{2(1-N)} a_1^\dagger a_1 - q_2^{2(1-N)} a_2^\dagger a_2\right] \end{aligned} \tag{23}$$

Using algebra 20, we can check that these generators satisfy the algebras 13.

4. Conclusions

In this paper we have constructed generators of of $SU_q(2)$ algebra in terms of q-deformed multimode boson oscillators which involves q- interference between oscillators of different modes (2.11) and in terms of q- deformed multimode boson oscillators in which each oscillator mode has its own deformation parameter (3.9). These realisations of $SU_q(2)$ algebra in terms of q- deformed multimode boson oscillators have subsequently found applications in the $SU_q(2)$ rotator model, as for example, rotational spectra of diatomic molecules.

Acknowledgment

The authors would like to thank Prof. Dao Vong Duc for helpful discussions.

References

[1] R. J. Finkelstein, *Lett. in Math. Phys.* **34**, (1996), 169.
[2] R. N. Mohapatra, *Phys. Lett. B* **242** (1990), 407.
[3] Y. Yang, and Z. Yu, *Mod. Phys. Lett. A* **8** (1993), 3025.
[4] O.W. Greenberg, *Phys. Rev. Lett.* **64**, 705.
[5] D. Bonatsos and A. Klein, *Nucl. Phys. A* **469** (1987), 253.

Journal of Physics: Conference Series **537** (2014) 012011 doi:10.1088/1742-6596/537/1/012011

The Equation of Backward Diffusion and Negative Diffusivity

Vu Ba Dung and Dinh Van Thien

Hanoi University of Mining and Geology, Dong Ngac, Tu Liem, Ha Noi, Vietnam

E-mail: vubazung305@gmail.com

Abstract. Up to now, all the diffusion theories have shown that the direction of diffusion flux is from high concentration area to lower concentration area (forward diffusion) and diffusivity is positive. In some research on simultaneous diffusion of boron and point defects in Si, the results show that the diffusivity of interstitials can be negative and diffusion process can be backward diffusion. In backward diffusion process, the direction of diffusion flux is from low concentration area to higher concentration area. The phenomenon of backward diffusion and negative diffusivity should be explained. Based on thermodynamic theory, the equation of backward diffusion and negative diffusivity are presented and discussed.

1. Introduction

The diffusion is elementary and universal process in natural. The rate law of diffusion had been formulated by A. Fick in which the diffusion flux is proportional to the gradient of the concentration and diffusion flux goes from regions of higher concentration to regions of lower concentration. The some piece of researches of us [1-5] showed that in the simultaneous diffusion of boron and point defect in silicon, the diffusivities of interstitial could be negative and diffusion process of interstitial or vacancy could be backward diffusion in silicon. In this paper, some problems of negative diffusivity, backward diffusion process, backward diffusion equation and solution of backward diffusion equation are presented and discussed.

2. Fick's laws of diffusion and forward diffusion process

Fick's first law postulates that the flux goes from regions of high concentration to regions of low concentration (forward diffusion) [6-9]:

$$J = -D\frac{\partial C}{\partial x} \tag{1}$$

in which J is the diffusion flux, D is the diffusivity, C is concentration and x is the position. Fick's second law predicts how diffusion causes the concentration to change with time t [6-9]

$$\frac{\partial C}{\partial t} = D\frac{\partial^2 C}{\partial x^2} \tag{2}$$

In equation (1) and (2) diffusivity D is positive and depends on temperature T, viscosity of fluid or gas and the size of the particles r according to the Stokes-Einstein relation [6-9]

$$D = \frac{kT}{6\pi\eta r} \tag{3}$$

The diffusivity in solids at different temperatures is found to be well predicted by Arrhenius equation [7]:

$$D = D_o exp(-\frac{E_a}{RT}) \tag{4}$$

in which D_o is the maximum diffusivity and E_a is the activation energy. Fick's laws have shown that the direction of diffusion fluxes are always the same direction of concentration gradient and the diffusivities are always positive sign.

3. The backward diffusion equation and negative diffusivity

Applying the thermodynamics theory on transport process could find the general equation of diffusion and general formula of diffusivity. Assume C_1 is deferent to C_2 and the thermal velocity u_2 and u_2 are not equal [5]

$$u_1 = u \tag{5}$$

$$u_2 = u_1 - nu \tag{6}$$

$$C_1 = C - \frac{\Delta C}{2} \tag{7}$$

$$C_2 = C + \frac{\Delta C}{2} \tag{8}$$

in which n is natural number and $\Delta C = C_2 - C_1 > 0$. The flux of molecules is determined by

$$J = -D\frac{\partial C}{\partial x} + wC \tag{9}$$

in which D and w are determined by expressions:

$$D = \frac{(2-n)\sqrt{3}u\lambda}{6} \tag{10}$$

$$w = \frac{\sqrt{3}nu}{3} \tag{11}$$

in which is characteristic length. Equation (9) is the general transport equation, equation (10) is the general diffusivity and equation (11) is the advection velocity. We could set up

$$J_{di} = -D\frac{\partial C}{\partial x} \tag{12}$$

$$J_{ad} = wC \tag{13}$$

in which J_{di} is diffusion flux and J_{ad} is advection flux. And equation (9) becomes:

$$J = J_{di} + J_{ad} \tag{14}$$

Transport flux J in equation (14) includes diffusion flux and advection flux.
The general transport equation (9) is also called advection-diffusion equation. This equation showed that the nature of transport process is depends on the difference between the thermal agitation velocity of molecules in high concentration areas and in low concentration areas.
i) If the thermal velocity of molecules in low concentration and in high concentration are equal ($u_1 = u_2$ or n = 0), so diffusivity is positive ($D > 0$) and the advection velocity is vanish (w = 0), the transport is only the diffusion process (the same Fick law - forward diffusion).
ii) If $u_2 < u_1 < 3u_2$ (0 < n < 2), so D > 0 and w > 0, the transport process includes the forward diffusion and the convection.
iii) If $u_1 = 3u_2$ (n=2), so D = 0 and w > 0, the transport process is only convection. Although there is difference of concentration in other areas, there is not diffusion process.
iv) If $u_1 > 3u_2$ (n > 2), so D < 0 and w > 0, the transport process includes the diffusion process with negative diffusivity and the convection.
When diffusivity is negative, the diffusion flux goes from the low concentration areas to high concentration areas. This diffusion process is contrary to Fick's law, so its called backward diffusion process. However, the backward diffusion process is different from forward diffusion process, both of them are described by the similar Fick's equations (1) and (2). If the diffusivity is positive, equation (1) and (2) are forward diffusion equations (FDE) [13-20]. If the diffusivity is negative, equation (1) and (2) are backward diffusion equations (BDE). In other word, process of backward diffusion is described by equation:

$$\frac{\partial C}{\partial t} = D\frac{\partial^2 C}{\partial x^2}(D < 0) \tag{15}$$

D is negative, so we can set up:

$$D = -D_p \tag{16}$$

in which D_p is positive. Substituting equation (16) into equation (15) we have:

$$\frac{\partial C}{\partial t} = -D_p\frac{\partial^2 C}{\partial x^2} \tag{17}$$

The theories of partial differential equation showed equation (17) have no solution. But we could find solution for equation (17) with some boundary conditions and initial conditions. We

Journal of Physics: Conference Series **537** (2014) 012011 doi:10.1088/1742-6596/537/1/012011

assume that the diffusion process occur one-dimensional (0 < x < L) with boundary conditions and initial conditions are:

$$C(0,t) = C(L,t) = C_0 \tag{18}$$

$$C(x,0) = f(x) \tag{19}$$

Solution for equation (17) with boundary and initial conditions (18) and (19) is given by:

$$C(x,t) = C_1(x,t) + C_2(x) + C_3(x) \tag{20}$$

Substituting equation (20) into equation (17) we have:

$$\frac{\partial C_1}{\partial t} + D_p\frac{\partial^2 C_1}{\partial x^2} + D_p\frac{\partial^2 C_2}{\partial x^2} + D_p\frac{\partial^2 C_3}{\partial x^2} = 0 \tag{21}$$

The function $C_2(x)$ is solution of equation:

$$D_p\frac{\partial^2 C_2}{\partial x^2} = 0 \tag{22}$$

with boundary condition:

$$C(0) = C_0 \tag{23}$$

$$C(L) = 0 \tag{24}$$

and function $C_2(x)$ is found:

$$C_2(x) = C_0(1 - \frac{x}{L}) \tag{25}$$

The function $C_3(x)$ is solution of equation:

$$D_p\frac{\partial^2 C_3}{\partial x^2} = 0 \tag{26}$$

with boundary condition:

$$C(0) = 0 \tag{27}$$

$$C(L) = C_0 \tag{28}$$

Journal of Physics: Conference Series **537** (2014) 012011 doi:10.1088/1742-6596/537/1/012011

and solution $C_3(x)$ is found:

$$C_3(x) = C_0\frac{x}{L} \tag{29}$$

The function $C_1(x)$ is solution of equation:

$$\frac{\partial C_1(x,t)}{\partial t} + D_p\frac{\partial^2 C_1(x,t)}{\partial x^2} = 0 \tag{30}$$

with boundary conditions and initial condition:

$$C_1(0,t) = C_1(L,t) = 0 \tag{31}$$

$$C_1(x,0) = f(x) - C_2 - C_3(x) = F(x) \tag{32}$$

the solution of equation (34) is:

$$C_1(x,t) = B.e^{(\frac{n\pi}{L})^2 D_p t}.sin\frac{n\pi}{L}x \tag{33}$$

where (n = 0, 1, 2) and factor B is determined by:

$$B = \frac{2}{L}\int_0^L F(\xi)sin\frac{n\pi}{L}\xi d\xi \tag{34}$$

The particular solution of backward diffusion equation (20) is:

$$C_n(x,t) = C_o + (\frac{2}{L}\int_0^L F(\xi)sin\frac{n\pi}{L}\xi d\xi)exp(-\frac{n^2\pi^2}{L^2}Dt)sin\frac{n\pi}{L}x \tag{35}$$

Example, the boundary and initial condition is given by:

$$C(0,t) = C(L,t) = C_o \tag{36}$$

$$C(x,0) = f(x) = C_o + C_s.sin\frac{n\pi}{L}x \tag{37}$$

The particular solution of backward diffusion is:

$$C_n(x,t) = C_o + C_s exp(-\frac{n^2\pi^2}{L^2}Dt)sin\frac{n\pi}{L}x \tag{38}$$

We can plot the graph of function (38) for backward diffusion of self - interstitial defects in silicon. In fig.1 there are graphs of the particular solution of backward diffusion equation with diffusivity of interstitial in silicon $D = -10^{-12}cm^2s^{-1}$, $C_o = 5.10^{14}cm^{-3}$, $C_s = 10^{15}cm^{-3}$ and $L = 2.10^{-4}$ cm [1-3]. Line 1 is graphs of the particular solution (38) with t = 0 and n = 1. Line 2 is graphs of the particular solution (40) with t = t_1 = 300 s and n = 2. Line 3 is graphs of the particular solution (38) with t = t_1 = 300 s and n = 3. However, there is a problem with the general solution of backward diffusion equation (38) that is C(x, t) becomes extreme when n is great value.

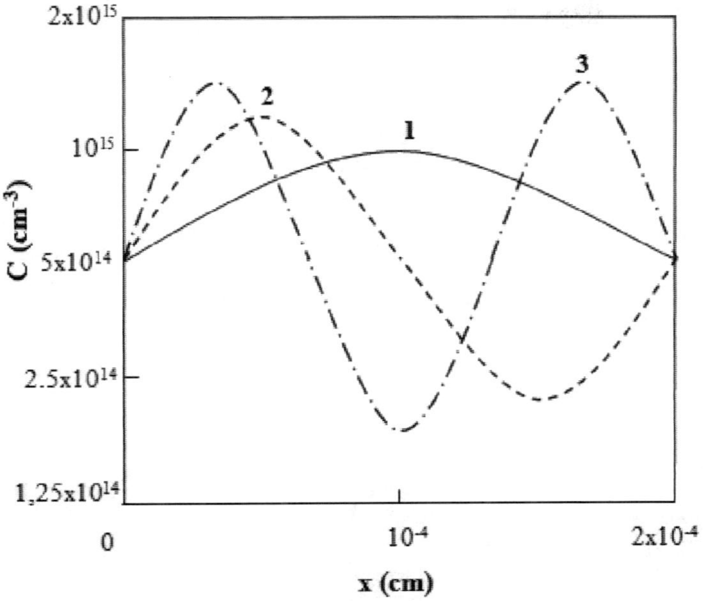

Figure 1. The graph of particular solution of BDE

4. Conclusion

The backward diffusion process and negative diffusivity is contrary to Ficks law, but these are explained by the theory of thermodynamics:

- Cause of backward diffusion and negative diffusivity is the difference between the thermal velocity of molecules in low concentration areas and the thermal velocity of molecules in high concentration areas.
- The backward diffusion process is described by an equation similar to Ficks equation.
- The backward diffusion equation could have particular solutions.
- Some problems of the particular solutions of the backward diffusion equation should be studied more in the future.

References

[1] Vu Ba Dung and Dao Khac An, *Proc. of IWOMS'99*(1999) 529-532.
[2] Vu Ba Dung and Dao Khac An, *Defect and Diffusion Forum* **194**(2001) 647- 652.
[3] V.B.Dung, D.K.An and P.A.Tuan and N.V. Truong, *Defect and Diffusion Forum*, **258**(2007) 32-38.
[4] G. Gilboa, N. Sochen and Y. Zeevi, *Journal of Mathematical Imaging and Vision* **20** (2004) 121-131.
[5] Vu Ba Dung at, *Conf. Theor. Phys* **37** (2012) 163-167
[6] A. Fick, *Phil. Mag* **10**(1855) 30.
[7] J. Philibert, *Diffusion Fundamentals 2* (2006) 1-10
[8] D. Brogioli and A. Vailati, *Phys. Rev.* **E 63** (2001) 012105/1-4.
[9] K. Olh, *Per. Polyt. Chem. Eng*, **49**(2005) 91-101.
[10] P. Perrot *A to Z of Thermodynamic* Oxford University Press (1998).
[11] P.Garbaczewski *Acta. Phys. Pol.* **E39** (2008) 1087-1101.
[12] G. Kuiken *Thermodynamics of Irreversible Processes*, Wiley (1994).
[13] I. Mller and W. Weiss *The European Physical Journal H* **37** (2012) 139-236.
[14] T. Myint-U, L. Debnath *Linear Partial Differential Equations for Scientists and Engineers* (2006).
[15] E. Zachmanoglou, D. Thoe, *Introduction to Partial Differential Equations With Applications* (1993).
[16] L.C.Evans and M. Portilheiro, *Math. Models Method Appl. Sci.* **14** (2009) 341-351.
[17] J. Jost *Partial Differential Equations* New York: Springer-Verlag (2002).

Journal of Physics: Conference Series **537** (2014) 012011 doi:10.1088/1742-6596/537/1/012011

[18] Y. Pinchover *An Introduction to Partial Differential Equations,* Cambridge Uni. Press (2005).
[19] P. Solin. *Partial Differential Equations and the Finite Element Method* J. Wiley and Sons (2005).
[20] S. Cardanobile, *Parabolic systems with coupled boundary conditions* **247** (2009) 12291248.

Journal of Physics: Conference Series **537** (2014) 012012 doi:10.1088/1742-6596/537/1/012012

Study Chaotic Behavior of a 3 Body Systems: Simple Application to Earth-Sun-Moon like System

N T T Nguyet, T T T Van, N V Thanh and N A Viet

Institute of Physics, Vietnam Academy of Science and Technology, 10 Dao Tan, Ba Dinh, Ha Noi, Vietnam

E-mail: conguyetdhhp1968@gmail.com

Abstract. It is well known that three-body systems, in general will exhibit chaotic behavior. In this work, we study the case of the simple restricted planar three-body problem, and its application to the Earth-Sun-Moon like system. Earth-like and Moon-like have a small mass relative to the Sun-like object. The Moon-like is affected both centrifugal and Coriolis forces, and it would not be able to escape from the Hill disc. The trajectories of the Moon-like was viewed in a rotating frame which fixes the two more massive bodies Sun-like and Earth-like, so able reduced to a simple two-degrees of freedom system. It is possible to construct a more generalized model for investigation the chaotic behavior of tide forces on Earth also discussed.

1. Introduction

Changes in the sea level are very well known as tides but the investigation it accurately is not easy because the Earth-Sun-Moon like system is a three-body problem, in general will exhibit chaotic behavior. The motion of the Moon-like around the Earth-like can be considered in first approximation as a Keplerian motion perturbed by the action of the distant Sun. The perturbation here is more important than in the planetary case and the problem was the object of major works since Newton himself and then the others later. In this first approximation the mass of the Moon-like is supposed to be zero, Poincarï£¡, followed by Birkhoff , developed the so-called restricted problem, where the motion of the two massive planets, being unperturbed by the zero mass body, is Keplerian. In this work in the frame of restricted problem we investigate the chaotic behavior of Moon-like object. This will be useful in the further understanding and analyzing our collected modern moon tide data of Hon-dau (Batshaw in the Newton' book Principia) tide observation station, where Newton took as an example to develop his famous gravity law.

2. The model

We treat the Sun-Earth-Moon like system as a three-body restricted problem in plane. In this case when the motion is circular and the small mass body (Moon-like) like body (M) in the same plane of Sun-like (S) and Earth-like (E) bodies. The problem can be studied in a rotating frame which fixes the two massive bodies (Sun-like and Earth-like). Assume the Earth-like object is fixed in the center 0, Sun-like object is located in the 0y direction of the rotating coordinate system (x0y), and Moon-like object is moving around the Earth-like body (see Fig.1).

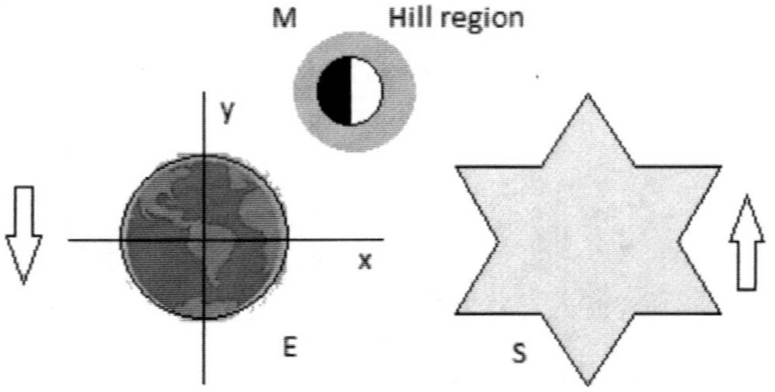

Figure 1. The Sun-Earth-Moon like system in a rotating frame.

In this rotating frame, the trajectory of the Moon-like object is defined by an autonomous Hamiltonian with two degrees of freedom, and the Jacobi constants. We have

$$\begin{aligned} x''(t) &= 2y'(t) + x(t) - \frac{\mu_1(\mu_2+x(t))}{\left[(\mu_2+x(t))^2+y(t)^2\right]^{3/2}} \\ &- \frac{\mu_2(-\mu_1+x(t))}{\left[(-\mu_1+x(t))^2+y(t)^2\right]^{3/2}}, \\ y''(t) &= 2y(t) - 2x'(t) - \frac{\mu_1 y(t)}{\left[(\mu_2+x(t))^2+y(t)^2\right]^{3/2}} \\ &- \frac{\mu_2 y(t)}{\left[(-\mu_1+x(t))^2+y(t)^2\right]^{3/2}}. \end{aligned} \tag{1}$$

where μ_i are the reduced masses of the Earth-like and Moon-like with Sun-like object, while x and y are the coordinates in the rotating system, the prime sign means derivative on time $\partial/\partial t$. The small body is subjected to both centrifugal and Coriolis forces. The masses are in the unit of the Sun-like object M_S, and the coordinates are taken in the units of distance between the Sun-like and Earth-like objects $R = R_E - R_S$, and the time unit is the period (year) of Earth-like body circling around the Sun-like object. The initial conditions are:

$$\begin{aligned} x(t=0) &= x_0, \\ y(t=0) &= y_0. \end{aligned} \tag{2}$$

for coordinates, and

$$\begin{aligned} x'(t=0) &= \nu_{x0}, \\ y'(t=0) &= \nu_{y0}. \end{aligned} \tag{3}$$

for initial velocities, correspondingly.
Actually the mass of Moon-like object is very small, so it is supposed to be zero$\mu_2 = \mu_M = 0$. The mass of the Earth-like body is also small in comparing with the Sun-like object $\mu = \mu_1 = \mu_E \approx M_E/M_S \lll 1$. We plot the trajectories of the Moon-like object with several sets of the initial conditions and $\mu = 3.10^{-6}$ like in the Sun-Earth system in the Fig. 2. The chaos behavior of trajectories of Moon-like object can be seen quite clearly.

2.1. The Hill region

The projections on the configuration plane of the constant energy surfaces define regions of possible motions, called the Hill regions. It occurs when, the Jacobi constant being negative and big enough. For simplicity, we consider the Moon-like object as a zero body mass turning around the more massive body (the Earth-like), moves in a component of the Hill region which

Journal of Physics: Conference Series **537** (2014) 012012 doi:10.1088/1742-6596/537/1/012012

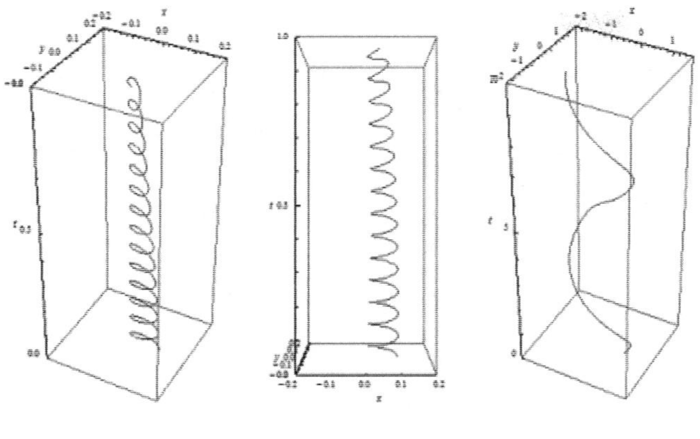

Figure 2. Trajectories of Moon-like object with chaos behavior.

is expressed by a disc with gray color in the Fig. 3. This fact already implies Hill's rigorous stability result: for all times such a Moon-like body would not be able to escape from this disc.

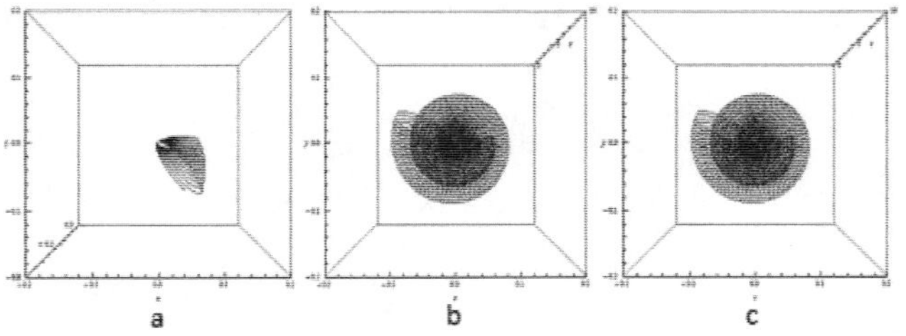

Figure 3. Hill regions of the Moon-like object.

2.2. Lagrange points

In some case, the Moon-like object appears to orbit around points other than the Earth-like and Sun-like objects. These positions are called "Lagrange points", are the equilibrium states in the co-rotating system. An example of this point with its chaotic behavior is shown in the

2.3. Tide and tide potential and force

Changes in the see level on the Earth are very well known as tides but the explanation is not easy as the simple statement that they are caused by gravity of the Moon. Beside that the Sun also gives an important contribution to the tides.
The magnitude of the potential and force attracting two bodies with masses m and M each other in the distance is given by the Newton's law of gravity with the gravitational constant G is given by $V(r) = GmM/r$ and $F(r) = GmM/r^2$, so the gravitational potential (in the unit mass) from the Moon-like body in the center O of the Earth-like object is

$$V_M(0) \quad = \quad \frac{GM_M}{r^2{}_M}, \tag{4}$$

where r_M is the distance between Earth-like and Moon-like objects.
Assume the distances between Sun or Earth like objects and the Moon-like object are much large than the Moon-Earth-like separation , the Moon-tide force in the surface of Earth-like object in lowest order of approximation can be considered its value at its center

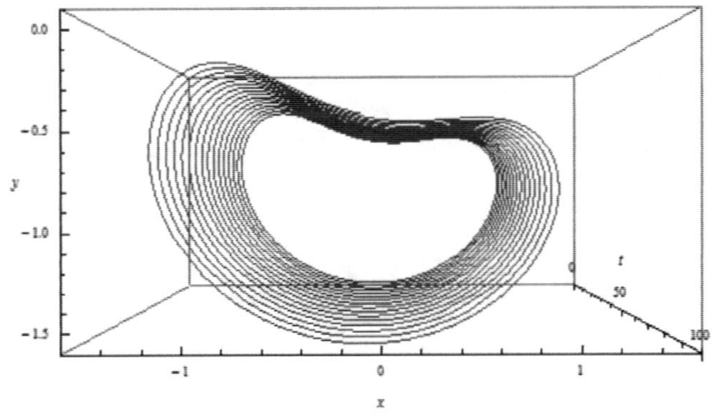

Figure 4. Example of Lagrange points.

$$F_M(0) = \frac{GM_M}{r_M^2}. \tag{5}$$

Because the distance between Moon and Earth like objects exhibit chaotic behavior so the values of tide potential and force also are chaos.

The Earth's axial tilt is 23.4o and the Moon's inclination to elliptic is 5.1o, but for simplicity we assume both are zero for the system of Earth-like and Moon-like, that the Earth-like's rotational axis is orthogonal to the orbital plane and the Moon-like orbits around the Earth-like above the Earth-like's equator, also assume that the hole Earth-like is covered with water.

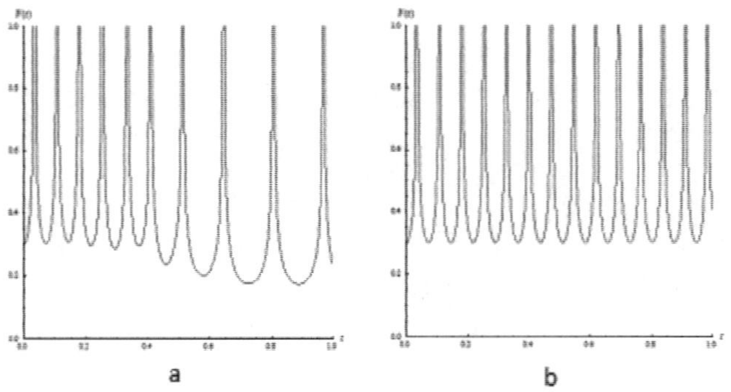

Figure 5. Chaotic behavior of the tide potential in the Earth like body from Moon and Sun like objects.

3. Discussions

In this work, chaotic behavior of a three body problem: the Earth-Sun-Moon like system was investigated in the frame work of restricted problem. Taken the mass ratios as those of the real Sun-Earth-Moon system we investigated the trajectories and chaotic properties of the Moon-like object in rotating frame. We show that because the distance between Moon and Earth like objects exhibit chaotic behavior so the values of tide potential and force in the Earth like body also are chaos.

Journal of Physics: Conference Series **537** (2014) 012012 doi:10.1088/1742-6596/537/1/012012

References

[1] R. M. Canup. *Science*, 338(1052), 2012.
[2] H. Xue D. Xu. *Ocean Dynamics*, 2012.
[3] I. Franc. *WDS'12 Proc. Contr. Papers, P. III*, (98), 2012.
[4] M. Cuk and S. T. Stewart. *Science*, 338(1047), 2012.
[5] L. R. M. Maas. *J. Phys. Ocenog.*, 32(870), 2001.
[6] E. Zeleny. *Restricted Three-body in a Plane. Math, Proj.*, 2012.

IWTCP1 & NCTP38 IOP Publishing
Journal of Physics: Conference Series **537** (2014) 012013 doi:10.1088/1742-6596/537/1/012013

General Pade Effective Potential for Coulomb Problems in Condensed and Soft Matters

B L Quyen, D N Mai, N M Hoa, T T T Van, N L Hoai and N A Viet

Institute of Physics, Vietnam Academy of Science and Technology, 10 Dao Tan, Ba Dinh, Hanoi, Vietnam

E-mail: `builequyenytb@gmail.com`

Abstract. Effective potentials for finding the ground states and physical configurations have essential meaning in many Coulomb problems of condensed and soft matters. The ordinary n-Pade approximation potentials define as the ratio of Pi(r)/Pi+1(r), where Pi(r) are the polynomials of i-th order of charge separation r, give quite good fit and agreement of calculation results and experimental data for Coulomb problems, where screening effects are not important or exchange photons still are massless. In this work we consider a general Pade effective potential by included a factor of exponential form, which could give more accurate results also for above mentioned cases. This general Pade effective potentials with analytical expressions were useful to perform analytical calculations, estimations and to reduce the amount of computational time for future investigations in condensed and soft matter topics. For example of soft matter problems, we study the case of MS2 virus, the general Pade potential gives much more correct results comparing with ordinary Pade approximation.

1. Introduction

In theoretical physics, one frequently encounters power series expansions whichs do not converge or converge very slowly, and there are many methods for accelerating the convergence of these sequences and the subsequent evaluation of the limit of an infinite sequence. Among them, the Pade approximation provides a practical method for performing numerically the analytic continuation of function. The Pade approximation is a very simple and powerful alternative to polynomial approximations for analytic functions. It is known that the "best" approximation of a function by a rational function of given order - under this technique, the approximant's power series agrees with the power series of the function it is approximating. The technique was developed around 1890 by Henri Pade, but goes back to Georg Frobenius who introduced the idea and investigated the features of rational approximations of power series [1].

The Pade approximation provides such a method, we define the [n,m] Pade approximant to f(z) as the ratio of Pn(z) and Qm(z) [2]:

$$f^{[n,m]}(z) \equiv \frac{P_n(z)}{Q_m(z)} = f(z) + 0.(z^{n+m+1}),$$

$P_n(z)$ and $Q_n(z)$ are polyminals of degree n and m respectively, which has the same n+m first derivatives as f(z) at z=0. e.g.

$$P_n(z) = b_1 + b_2 z + ... + b_N z^n.$$

Journal of Physics: Conference Series **537** (2014) 012013 doi:10.1088/1742-6596/537/1/012013

The meaing of

$$\frac{P_n(z)}{Q_m(z)}$$

is that we can write:

$$f(z) = \underbrace{\frac{P_n(z)}{Q_m(z)} + d_{n+m+1}z^{n+m+1} + d_{n+m+z}z^{n+m+z} + ...,}_{0(z^{n+m+1})}$$

where the d's and also the coefficients of z the Pade approximate are functions of the coefficients in the Taylor series expansion.
From this point of view, the set of Pade approximants are a generalization of the Taylor series expansion - the [n, 0] approximants.
The Pade approximant often gives better approximation of the function than truncating its Taylor series, and it may still work where the Taylor series does not converge [15]. These techniques and concepts are found beside the benefit that is convergence acceleration (e.g. ϵ- algorithm), this method could be applicated to numerical solutions to partial differential equations ($\exp(At) \approx Q(At)^{-1}P(At)$), analytic continuation of power series (regions of convergence beyond a disk). It also includes study of orthogonal polys on interval (Pade denominators for Markov functions are orthogonal) and finding zeros/roots, poles/singularities (use zeros and poles of Pade approximants to predict - e.g. QD algorithm)[3].
All these interesting features, and the particular simplicity of the Pade approximation make it a very convenient tool for practical and physical applications. The range of applications to physical problems is very broad [16]. The first physical applications of the Pade approximation have been made in statistical mechanics[4]. Then there were many other application: in fluid [5], blasius problem [6]. In this work, we give particular emphasis to condensed and soft matter. We discuss various proofs of convergence, and show the approximation is particularly well suited for Coulomb problems. We finally review the various achievements of the approximation in the determination of Coulomb effective potential.
Very often, the equations describing a physical process are so complicated that the simplest way to obtain their solution, if not the only way, is to perform a power series expansion in some parameters. Furthermore, the physical values of the parameters may be such that this perturbation to the problem, i.e., it cannot be used quantitatively as such. However, the information is present in the coefficients of the perturbation series, and one may look for mathematical techniques that would be capable of treating this information in a convergent way. As above, we known that the Pade approximation use n+m+1 parameter, in this work, for simle, we just consider the case of m=n+1, so there are 2n+2 parameter here. There is a study about the modified pade approximation have been made, such as [7]. Now, we introduce a new modifies pade approximation, the general pade include a factor of exponential form, so there more one parameter in this extended part, it will make the convergence faster. The General Pade approximation is very well suited for as we shall see.

2. Method

2.1. Potential in one layer on the screening effect of a nearby ground-plane

As an example of how the approximation can be used, it is of interest to study briefly what can be considered as a commonly used Pade approximant in physics: the screening effect of a ground-plane on a two-dimensional system.
In a two-dimensional electron system (2DES), strong Coulomb interactions between electrons can lead to exotic phenomena such as the Wigner crystal state, the fractional quantum Hall

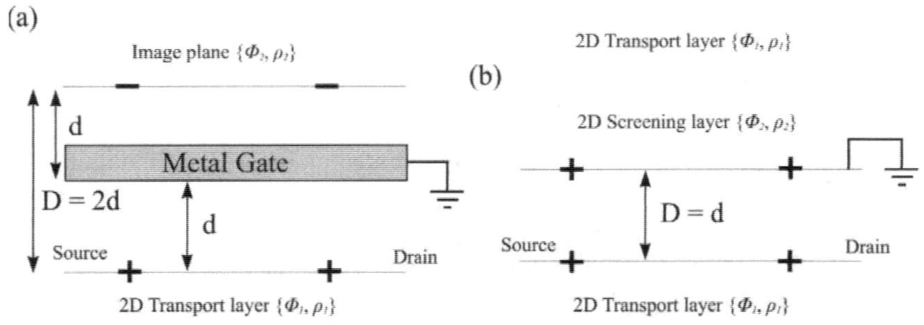

Figure 1. Schematics showing the two systems considered in this paper. The transport layer is screened by (a) a metal surface gate and (b) a second 2D system. In both cases the screening layer is separated by a distance d from the transport layer, and the transport (1) and screening (2) layers have independent potentials ϕ and charge densities ρ.

effect, and the anomalous 2D metallic state. So, the role of studying Coulomb interactions is very important.

We now begin considering the screening effect of a nearby ground-plane on a 2D system (transport layer) for two different coniňĄgurations. In the ïňĄrst, the ground-plane (i.e., screening layer) is a metal surface gate (see Fig. 2(a)) and in the second, the ground-plane is another 2D system (see Fig. 2(b)). In both cases the transport and screening layers are separated by a distance d. If we consider some positive external test charge ρ_{ext1} added to the transport layer, this leads to induced charge in both the transport layer ρ_{ind1} and in the screening layer ρ_{ind2}. Charge in one layer leads to a potential in the other via the interlayer Coulomb interaction[9]:

$$U(q) = \frac{1}{4\pi\varepsilon\sqrt{r^2 + D^2}}.$$

First, we considered some ordinary Pade approximations, $U(q)^{[1,2]}$, $U(q)^{[2,3]}$, $U(q)^{[3,4]}$ and also consider the Taylor series of (1). We see that the ordinary Pade approximation just accurate in a small range of r (Fig.2). Beside, we also consider the Taylor series of Eq.(*), it's the dashing black line, it's worse than ordinary Pade approximation.

Now, we will look for an analytical expression that can fits the wider range of r-values. Therefore, we have approximated the full effective potential by general Pade approximant. It's form is: $F = \frac{a_0}{r}e^{-b_0 r}$, where a_0, b_0 is determined after fitting the data to expression in Eq.(*), we found $a_0 = 0.0774905, b_0 = 15.8339.10^{-6}$.

The ordinary Pade approximant is known better than Taylor series, for this function we also use general Pade approximant, contained a factor of exponential, formed then compared with the cases above (Fig. 2). We see that the new Pade approximation is fitter for a larger range of r than other ordinary Pade approximations and ofcourse's Taylor series.

2.2. Two soft particles interaction

In another case in soft matter, we now consider consider the electrostatic interaction between two dissimilar spherical soft spheres 1 and 2 (figure 4). We denote by dl and d2 the thicknesses of the surface charge layers of spheres 1 and 2, respectively. Let the radius of the core of soft sphere 1 be and that for sphere 2 be a2 . We imagine that each surface layer is uniformly charged. Let Z1 and N1 , respectively, be the valence and density of the fixed charge layer of sphere 1 and Z2 and N2 be those for sphere 2.

For the special case of two similar soft spheres carrying Z1=Z2=Z, N1= N2= N, d1= d2= d so that $\rho_{fix1} = \rho_{fix2} = \rho_{fix}$. The interaction energy Vsp(H) between two similar soft spheres 1

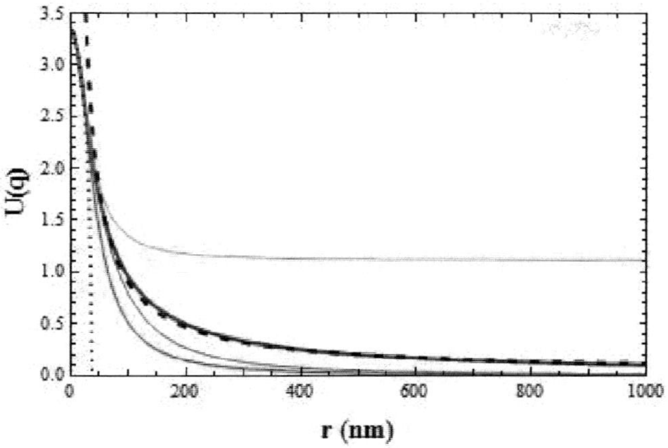

Figure 2. The potential in one layer is caused by the charge in the other layer, the blue, thick line is of Original function; the black, dashing, thick is of General Pade approximation, red: Pade [1,2]; green: Pade [2,3]; brown: Pade [3,4] and the black, dotted's one is for Taylor series.

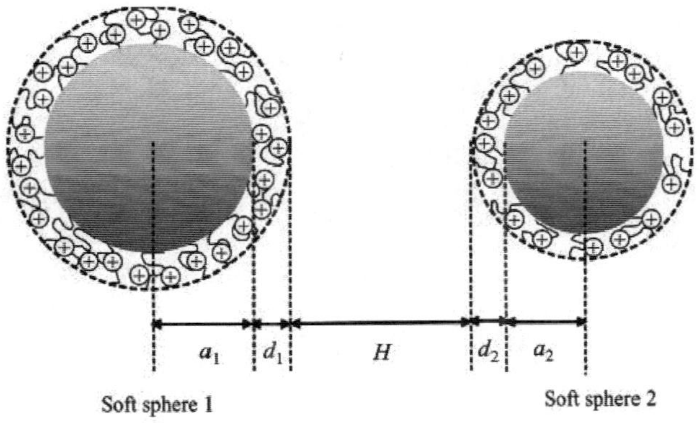

Figure 3. Interaction between two soft spheres.

and 2, separated by a ρ_{fix}, and a1= a2= a, equation (79) reduces to distance H between their surfaces is[3,4]:

$$V_{sp}(H) = \frac{2\pi a \rho_{fix}^2 \sinh^2(\kappa d)}{\varepsilon_r \varepsilon_0 \kappa^4} \ln\left(\frac{1}{1 - e^{-\kappa(H+2d)}}\right).$$

With a more complicated function, the approximants give much different. The Taylor series is worst of them (Fig. 4a), and the ordinay Pade approximants have more error than general Pade approximant (Fig. 4b). So in this case, the General Pade is more exactly than any others.

3. Conclusion

The ordinary Pade approximation is known that better than Taylor series, this work shown that the general Pade approximation by included a factor of exponential form give more accurate results than the ordinary Pade approximation, it fitted with a larger range, especialy in the complicated function. This new Pade approximation giving a simple approximant function formed of $F = \frac{a_0}{r} e^{-b_0 r}$, is useful to perform analytical calculations, estimations and to reduce the amount of computational time for future investigations in condensed and soft matter topics.

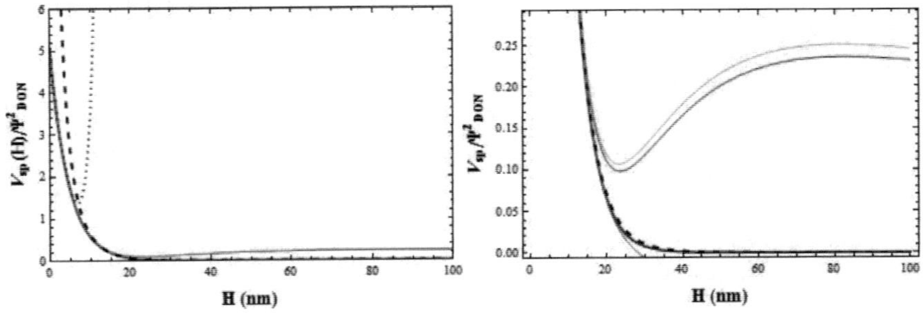

Figure 4. The interaction energy between two similar soft spheres, the blue, thick line is of Origin function; the black, dashing, thick is of General Pade approximation, red: Pade [1,2]; green: Pade [2,3]; brown: Pade [3,4] and the black, dotted's one is for Taylor series.

References

[1] B. Partoens W. Magnus A. F. Slachmuylders and F. M. Peeters. Pade apprimation exciton in nano wire. pages 2–14, 2005.

[2] Faiz Ahmad and Wafaa Alhasan Albarakati. Pade Application to solve the blasius problem. *Proc. Pakistan Acad. Sci.*, 44(1), 2007.

[3] J. L. Basdevant. The pade approximation and its physical application. 1971.

[4] C. Brezinski. History of Continued Fractions and Pade Approximants. *Springer-Verlag*, 1991.

[5] Ohshima H. Theory of Colloid and Interfacial Electric Phenomena. 12, 2006.

[6] Ohshima H. Theory of electrostatics and electrokinetics of soft particles. *Sci. Tech- nol. Adv. Mater.*, 10:063001–063013, 2009.

[7] Josef Kallrath. On rational function techniques and Pade approximants – An overview. 2002.

[8] A.P. Micolich A.R. Hamilton and O.P. Sushkov L.H. Ho. Ground-plane screening of Coulomb interactions in two-dimensional systems: How eïňĂectively can one two-dimensional system screen interactions in another? 2009.

[9] B.T.L. Quyen. Phisical modeling for virus. 2012.

[10] Marc Prevost and Tanguy Rivoal. Remainder Pade approximants for the exponential function.

[11] E. B. SaïňĂ. INTRODUCTION TO PADE APPROXIMANTS.

[12] N.T.L. Hoai H.P. Thao B.T.L. Quyen and N.A. Viet T. Huong Dang. Electric potential profile of spherical soft particle with point charged hardcore. *Proc. 37th Nat. Conf. Theo. Phys.*, 2012.

[13] A. Halder T. K. Sheel. Application of Pade approximation to problems of Fluid dynamics. *SUST Studies*, 4(1), 2002.

[14] E. B. Saff and R. S. Varga. On the zeros and poles of Pade Approximants to. *Springer-Verlag, Number. Math.*, (25, 1-14, 1975):1–14, 1975.

Journal of Physics: Conference Series **537** (2014) 012014 doi:10.1088/1742-6596/537/1/012014

Optical Trapping of Cold Neutral Atoms Using a Two-Color Evanescent Light Field Around a Carbon Nanotube

Do Thi Nga[1], Nguyen Ai Viet[1], Dao Thi Thuy Nga[2] and Nguyen Thi Phuong Lan[3]

[1]Institute of Physics, Vietnam Academy of Science and Technology, 10 Dao Tan, Ba Dinh, Ha Noi, Vietnam

[2]Han Thuyen High School, Bac Ninh City, Bac Ninh Province, Vietnam

[3]Hanoi Pedagogical University 2, Nguyen Van Linh Street, Xuan Hoa Ward, Phuc Yen Town, Vinh Phuc Province, Vietnam

E-mail: daothuyngaht@gmail.com

Abstract. We suggest a new schema of trapping cold atoms using a two-color evanescent light field around a carbon nanotube. The two light fields circularly polarized sending through a carbon nanotube generates an evanescent wave around this nanotube. By evanescent effect, the wave decays away from the nanotube producing a set of trapping minima of the total potential in the transverse plane as a ring around the nanotube. This schema allows capture of atoms to a cylindrical shell around the nanotube. We consider some possible boundary conditions leading to the non-trivial bound state solution. Our result will be compared to some recent trapping models and our previous trapping models.

1. Introduction

Recently, study properties of cold neutral trapped atoms have attracted much interest because of their significance in both fundamental research and potential technological applications. Cold and trapped radioactive atoms can be used in the fundamental symmetry experiments, including the experiments on nuclear β-decay, atomic parity non conservation and the search for parity and time-reversal violating electric dipole moment [1]. Trapping of ultracold atoms also gives an opportunity to study the collisional processes in cold atomic samples [2]. Trapped cold atoms can be used in the formation of cold molecules [3] and in studies of quantum statistical effects in atomic ensembles at low temperatures, such as the Bose-Einstein condensation (BEC). Since a cold molecules trap lifetime was approximately half a second, so the production of cold molecules opens up new ways of research in molecular spectroscopy.

There are different schemes for trapping and storing cold atoms [4], [5] such as optical trapping using the forces of electric dipole interaction between atoms and laser fields, magnetic trapping based on the use of the forces of magnetic dipole interaction, mixed magneto-optical trapping using simultaneous interaction between atoms and magnetic and laser fields, and also mixed gravito-optical and gravito-magnetic trapping. It has been shown experimentally that neutral atoms can be captured by an optical trap [6], based on the use of an evanescent wave around a thin silica fiber. The attractive potential created by the light is inversely proportional to the

Journal of Physics: Conference Series **537** (2014) 012014 doi:10.1088/1742-6596/537/1/012014

radius of the optical fiber. The condition for the existence of the trap is that the fiber diameter is about two times smaller than the light wavelength. Recently, a cold atom trap model using a charged single wall carbon nanotube (SWCNT) has been presented [7].
Based on these models, in the previous works, we have proposed a new trapping schema for the cold atoms using a single wall carbon nanotube [8]. In that model, for the existence of a stable bound state of cold atom, a strong electromagnetic field has been sent through the carbon nanotube. This field generates an evanescent wave around the carbon nanotube and creates an effective attractive potential. The consideration of some possible boundary conditions leads to this nontrivial bound state solution. There also is a comparison our result to the two most recent models concerning trapping of cold atoms by using a charged carbon nanotube and an optical fiber. It has been shown that the smaller the radius is, the stronger potential is created. The bound state in our model is more stable and resonalbe. New technology of nano materials is developed to optimize the trapping effect.
In this paper, a new design of optical trapping cold atom using a single wall carbon nanotube injected by two strong electromagnetic waves is proposed. This scheme gives strong effect on cold neutral atom, which need to be shown experimentally.

2. Two evanescent light field trap for cold neutral atom moving around a carbon nanotube

Consider an atom moving in a potential around the single wall carbon nanotube (SWCNT). If this potetial has a local minimum at a point outside the SWCNT, the atom can be trapped in the vicinity of this point. Our task is to create a potential with a trapping minimum sufficiently far from the SWCNT surface to make the effects of surface interaction and heating negligible. For this purpose, we use two light fields in the fundamental modes 1 and 2 with differing wavelengths λ_1 and λ_2, respectively. A schematic of our design is shown in Fig. 1.

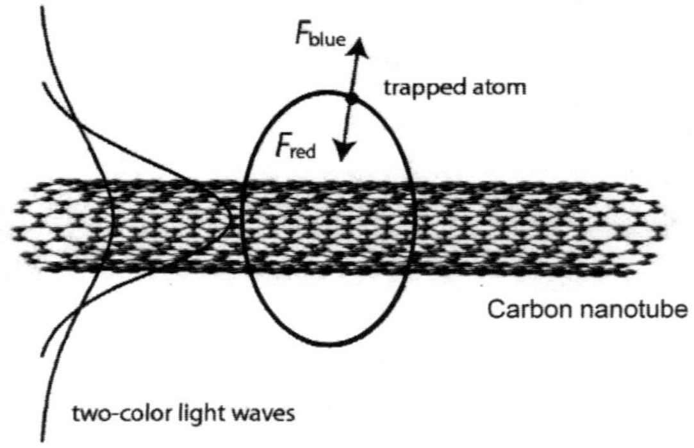

Figure 1. *Design of optical trapping cold atom by sending two electromagnetic waves through a carbon nanotube.*

The schema of our model is an analogy of the optical trap model [9], but the optical fiber is replaced by a SWCNT with radius R. The two electromagnetic waves with wavelengths λ_1 and λ_2 are sent into the SWCNT, the waveguide possibility of carbon nanotube is observed experimentally [10]. By evanescent effect, the wave decays away from the SWCNT surface producing an attractive potential for the neutral atom. The atom needs to be kept away from the SWCNT for a stable trapping. For a clear comparison with the optical trap model using a silica fiber [9] and single wall carbon nanotube, we use the same, red-detuned light λ_1 and blue-detuned light λ_2, as a source of electromagnetic waves.

We assume that the input light fields are circularly polarized to produce the cylindrically symmetric optical potential. The optical potential of the atom is cylindrically symmetric, that depends on the radial distance r from the atom to the carbon nanotube axis z, but not on two other cylindrical coordinates φ and z. The optical field generates an evanescent wave around the SWCNT to capture cold atoms. Assume that the time scale of atomic motion is much slower than the beating period of the two light fields, that is, the inverse of their frequency difference. Then, the optical potentials of the two fields add up to give the net optical potential

$$U_{\rm opt}(r) = U_1 + U_2 \tag{1}$$

The sign of the optical potential of each mode is controlled by the sign of the mode detuning. We choose a red detuning (the detuning of the optical frequency from the atomic frequency is negative) for the field in mode 1 and a blue detuning (the detuning of the optical frequency from the atomic frequency is positive) for the field in mode 2. This allows both attractive (red-detuned) and repulsive (blue-detuned) potentials to be created. In the linear-polarization approximation, the net optical potential outside SWCNT can be written as

$$U_{\rm opt}(r) = G_2 K_0^2(q_2 r) - G_1 K_0^2(q_1 r) \tag{2}$$

where G_1 and G_2 are positive coupling parameters for the interaction between the evanescent waves and the atom. They are proportional to the powers of the corresponding light fields. The parameter $q = 1/\Lambda$ is the inverse of the characteristic decay length Λ of the evanescent-wave field. The modified Bessel function $K_0(qr)$ is used to describe the spatial dependence of the amplitude of the field outside the carbon nanotube in the linear-polarization approximation.
The atom can have a stable bounded state only if $U_{\rm opt}(r)$ has a local minimum at distance $r = r_m$ outside of the carbon nanotube; that means the potential $U_{\rm opt}$ is attractive. The r_m is the solution of, $U'_{\rm opt}(r_m) = 0$, from this we find

$$F(r_m) = G \tag{3}$$

where

$$F(r) = \frac{K_0(q_1 r)K_1(q_1 r)}{K_0(q_2 r)K_1(q_2 r)} \tag{4}$$

and

$$G = \frac{G_2 q_2}{G_1 q_1} \tag{5}$$

Since $q_1 < q_2$, the function $F(r)$ is a monotonically increasing function of r. The local minimum point r_m exist only when $G > F(R)$. The requirement $r_m > R$ can be satisfied only when

$$\frac{G_2}{G_1} > \frac{q_1 K_0(q_1 R)K_1(q_1 R)}{q_2 K_0(q_2 R)K_1(q_2 R)} \tag{6}$$

Condition (6) means that there exists a trapping minimum outside the fiber if the intensity of the blue-detuned light is large enough or if the intensity of the red-detuned light is small enough (but not zero).
The atom can be prevented from coming too close to the SWCNT surface before being loaded into the trap if the net optical potential achieves a non-negative value at the fiber surface, that is, if $U_{\rm opt}(R) \geq 0$. This requirement will be satisfied if

$$\frac{G_2}{G_1} \geq \frac{K_0^2(q_1 R)}{K_0^2(q_2 R)} \tag{7}$$

When condition (7) is satisfied, condition (6) is also satisfied. To maximize the depth of the trapping minimum, we optimize the powers of the two field in such a way that condition (7) reduces to an equality, namely,

$$\frac{G_2}{G_1} = \frac{K_0^2(q_1 R)}{K_0^2(q_2 R)} \tag{8}$$

Due to the cylindrical symmetry of the total optical potential U_{opt}, the component L_z of the angular momentum of the atom is conserved. In the eigenstate problem, we have $L_z = \hbar m$, where m is an integer, called the rotational quantum number. The centrifugal potential of the atom is given by $U_{\text{cf}}(r) = \frac{(m^2-1/4)\hbar^2}{2Mr^2}$. The radial motion of the atom can be treated as the one-dimensional motion of a particle in the effective potential

$$U_{\text{eff}}(r) = \frac{(m^2-1/4)\hbar^2}{2Mr^2} + G_2 K_0^2(q_2 r) - G_1 K_0^2(q_1 r) \tag{9}$$

In our calculations, we use the SWCNT with (n, m)=(21,0) corresponding to the radius is $R = 0.83$ nm and set the parameters: $\lambda_1 = 1.2$ μm, $\Lambda_1 = 24.8$ nm, $\lambda_2 = 0.7$ μm, and $\Lambda_2 = 4.6$ nm. In the Fig. 3, it is observed that the existence of a deep local minimum of the effective optical potential outside carbon nanotube leads to the existence of captured states of the atom.

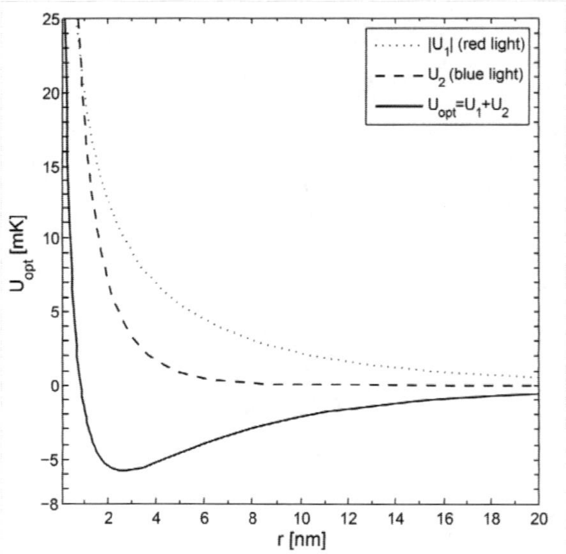

Figure 2. *The net optical potential outside the SWCNT (n, m)=(21,0) with the two evanescent waves λ_1 = 1.2 μm, λ_2 = 0.7 μm.*

3. Energy of the neutral cold trapped atoms in the effective potential around the a carbon nanotube

In previous section, we have been shown that the atom can be trapped by injecting two strong electromagnetic waves into a SWCNT. Therefore, the binding energy of the atom will be estimated by solving the Schrodinger equation.

The Schrodinger equation is written in the cylindrical coordinates (r,φ,z) in the form

$$[-\frac{\hbar^2}{2M}\nabla^2 - U_{\text{eff}}(r)]\,\Psi(r,\varphi,z) = E\,\Psi(r,\varphi,z) \tag{10}$$

where E is the energy of atom in the effective potential.
The general form of the wave function is used

$$\Psi(r,\varphi,z) = \frac{1}{\sqrt{r}} R(r)\, e^{im\varphi}\, e^{ikz} \tag{11}$$

substitute $\Psi(r,\varphi,z)$ into Schrodinger equation (10), we obtain the Schrodinger equation for center-of-mass mouvement of the atom in the presence of effective optical potential $U_{\text{eff}}(r)$ is

$$\frac{d^2R}{dr^2} + \frac{2\mu}{\hbar^2}[E - U_{\text{eff}}(r)]R = 0 \tag{12}$$

There is no analytic solution for equation (12), so an approxmation is made for the effective optical potential. Therefore, the potential could be written in the form $\frac{A}{r^2} - \frac{B}{r} + C$ A, B and C are the fitting parameters. The approximated effective potential is made with $A = 25.0$, $B = 19.3$ and $C = -2.1$. The solution of Eq. (12) with the approximated effective potential has the form [11]

$$E_n = C - B^2k^2\{2n + 1 + \sqrt{1 + 4Ak^2}\}^{-2},\ n = 0, 1, 2, ... \tag{13}$$

This expression gives the binding energy of cold neutral atom in the effective optical potential around a SWCNT. The experimental relization of bound states requires the binding energy of

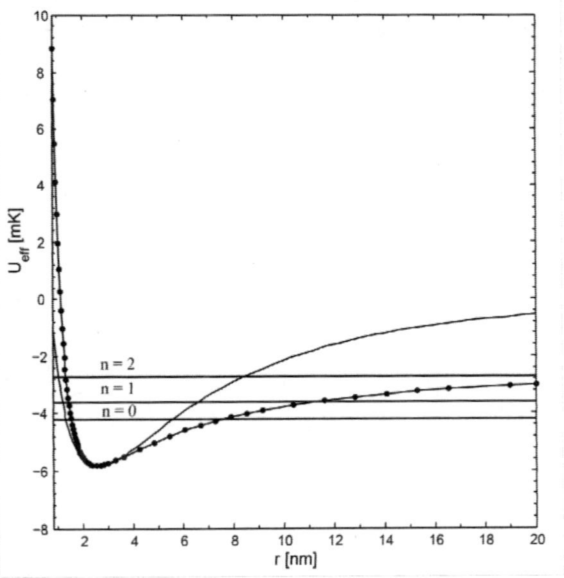

Figure 3. *Theorical effective potential (solid line) and approximate one (dotted line). Bound states for the first three levels (n = 0, 1, 2)of the radial motion of a cesium atom in the effective potential of a single wall carbon nanotube (m, n)=(21,0).*

atoms to be higher than their typical thermal kinetic energy. In Fig. 3, we plot the first three levels of the radial motion of the atom in the effective potential U_{eff} with the rotational quantum number $m = 0$. The energy of the ground state is $E_0 \approx -4.2$ mK. Since the binding energy of an atom in the ground state is $E_b = U_{\text{opt}}(r = \infty) - E_0 \approx 4.2$ mK. Thus our design can trap cesium atoms around the SWCNT at a temperature of less than 4.2 mK.

4. Conclusion

It has been shown that the neutral cold atoms moving near a SWCNT can be captured by sending two evanescent light field through it. It's a new model for trapping cold atoms, based on the ideas of some recent experimental and theoretical model [8]-[9]. In comparing with our previous trapping design [8], the ground state energy is much more smaller, trapping effect is stronger. Cold atom can be trapped and store to be study easily. This is a new design of optical trap which give strong effect on cold neutral atom, which need to be shown experimentally.

Acknowledgment

This research was financially supported by the National Foundation for Science and Technology Development (NAFOSTED) of Vietnam through Grant No. 103.01-2013.25.

References

[1] J. A. Behr et al., Phys. Rev. Lett. 79 (1997) 375.
[2] G. M. Bruun and K. Burnett, Phys. Rev. A 58 (1998).
[3] A. Fioretti, D. Comparat, A. Crubellier, O. Dulieu, F. Masnou-Seeuws and P. Pillet, Phys. Rev. Lett. 80 (1998) 4402-4405.
[4] F. L. Kien, V. I. Balykin and K. Hakuta, Phys. Rev. A 74 (2006) 033412.
[5] G. Petrov, S. Machluf, S. Younis, R. Macaluso, T. David, B. Hadad, Y. Japha, M. Keil, E. Joselevich and R. Folman, Phys. Rev. A 79 (2009) 043403.
[6] V. I. Balykin, K. Hakuta, Fam Le Kien, J. Q. Liang and M. Morinaga, Phys. Rev. A 70 (2004) 011401(R).
[7] P. R. Giri, Int. J. Theor. Phys. 47 (2008) 1776-1783.
[8] T. A. Chu, D. T. Nga, T. T. Thao, V. Thanh Ngo, N. A. Viet, Mod. Phys. Lett. B, vol. 25, 2011, pp. 979-985.
[9] Fam Le Kien, V. I. Balykin and K. Hakuta, Phys. Rev. A, vol. 70, 063403 (2004).
[10] M. V. Shuba et al., Phys. Rev. B 76 (2007) 155407.
[11] L. D. Landau and E. M. Lifshitz, Quantum Mechanics (Pergamon Press, London, 1959).

Journal of Physics: Conference Series **537** (2014) 012015 doi:10.1088/1742-6596/537/1/012015

Some Decays of Neutral Higgs Bosons in the NMSSM

Nguyen Chinh Cuong, Do Thi Thu Trang and Nguyen Thi Phuong Thuy

Hanoi National University of Education, 136 Xuan Thuy, Cau Giay, Ha Noi, Vietnam

E-mail: cuongnc@hnue.edu.vn

Abstract. To solve the μ problem of the Minimal Supersymmetric Standard Model (MSSM), a single field S is added to build the Next Minimal Supersymmetric Standard Model (NMSSM). Vacuum enlarged with non-zero vevs of the neutral-even CP is the combination of H_u, H_d and S. In the NMSSM, the higgs sector is increased to 7 higgs (compared with 5 higgs in the MSSM), including three higgs which are even-CP $h_{1,2,3}$($m_{h1} < m_{h2} < m_{h3}$), two higgs which are odd-CP $a_{1,2}(m_{a_1} < m_{a_2}$) and a couple of charged higgs $H^{\pm}$. The decays higgs into higgs is one of the remarkable new points of the NMSSM. In this paper we study some decays of neutral Higgs bosons. The numerical results are also presented together with evaluations.

1. Introduction

The simplest version of supersymmetry is the Minimal Supersymmetric Standard Model (MSSM). This version is limited by two problems: the μ and the hierarchy [1, 3, 4, 7]. The simple supersymmetry, which is beyond the MSSM, is the Next Minimal Supersymmetric Standard Model (NMSSM). The special characteristic of Higgs boson in the NMSSM is the decay of Higgs into Higgs. It is remarkable that the lightest state a1 of the odd-CP Higgs can play the role of a pseudo-goldstone, which has a small mass and can lead to the predominated decay of the even-CP $h \to a_1a_1$ [2]. The even-CP Higgs and the heavy odd-CP Higgs can be generated at LEP in $e^+e^- \to ha$, but they may not be discovered because the dominant h decay were not searched for. There are different ways to make the mass of Higgs boson increased in the MSSM and in the beyond MSSM. One simple way is to study the beyond singlet of the MSSM which contains one term $\lambda\hat{S}\hat{H}_u\hat{H}_d$ in the super-potential, this is the term that contributes $\lambda^2v^2\sin^2 2\beta$ at v = 174 GeV to the squared mass of even-CP Higgs [10] and therefore, it can make the mass of Higgs boson increased over the limit of independent decay state. It should be noted that this contribution is maximum with $\tan\beta \sim 1$. Thus, a condition in which the lightest odd-CP Higgs with its mass is under $2m_b$, the two lightest even-CP Higgs boson and the charged Higgs boson can be found in the MSSM. All of them can be generated at LEP and they are now being searched for.

The neutral Higgs sector in the NMSSM includes the following states: three even- CP and two odd-CP. Many analys is on Higgs sector in the NMSSM [5] have shown that, in the specific physical state of the even-CP Higgs, there is a strong mix between the doublet state and the singlet SU(2) with the reduction in the interaction of gauge boson. The study on light Higgs

contributes to the discovery of one or more Higgs states at LEP, at LHC [5] and at large energy accelerators.
The NMSSM is established from the MSSM when the MSSM is added with a gauge chiral single superfield $\hat{S}$, which includes the renormalized superfield interaction and the soft supersymmetry breaking term L_{soft}. In the NMSSM, the terms of the super-potential W_{higgs} are dependent on superfield Higgs $\hat{H}_d$, $\hat{H}_u$ and $\hat{S}$ (here, we follow the SLHA2 regulations, however $\hat{H}_u$ is also written as $\hat{H}_d$ and $\hat{H}_d$ is also written as $\hat{H}_1$):

$$\mathrm{W}_{Higgs} = \left(\mu + \lambda\hat{S}\right)\hat{H}_u.\hat{H}_d + \xi_F\hat{S} + \mu'\hat{S}^2 + \frac{\kappa}{3}\hat{S}^3. \tag{1}$$

with: λ, κ is the non-dimension coupling Yukawa, μ, μ' is the supersymmetry mass, ξ_F is the square supersymmetry mass parameter.
From (1), Yuakawa interaction of quark and lepton superfield are added to

$$\mathrm{W}_{Yukawa} = h_u\hat{H}_u.\hat{Q}\hat{U}_R^c + h_d\hat{H}_d.\hat{Q}\hat{D}_R^c + h_e\hat{H}_d.\hat{L}\hat{E}_R^c. \tag{2}$$

Here, the Yukawa interaction h_u, h_d, h_e and the superfields $\hat{Q}, \hat{U}_R^c, \hat{D}_R^c, \hat{L}, \hat{E}_R^c$ are the matrices and vectors in the corresponding spaces.
The soft breaking supersymmetry sector is regulated in SLHA2:

$$\begin{aligned}-L_{\rm soft} &= m_{Hu}^2|H_u|^2 + m_{Hd}^2|H_d|^2 + m_s^2|S|^2 + m_Q^2\left|Q^2\right| + m_U^2\left|U_R^2\right| \\ &+m_D^2\left|D_R^2\right| + m_L^2\left|L^2\right| + m_E^2\left|E_R^2\right| \\ &+(h_uA_uQ.H_uU_R^c - h_dA_dQ.H_dD_R^c - h_eA_eL.H_dE_R^c \\ &+\lambda A_\lambda H_u.H_dS + \frac{1}{3}\kappa A_\kappa S^3 + m_3^2H_u.H_d + m_s^{2'}S^2 + \xi_sS + hc).\end{aligned} \tag{3}$$

(definition $m_3^2 = B\mu$,$m_s^{'2} = B'\mu'$)
In the super-potential (1) we have supersymmetry parameters μ, μ' and ξ_F (the soft supersymmetry interactions break the parameters $m_3^2, m_s^{'2}$ and parameter ξ_S in (3)), however, some terms are not eliminated in some different solutions for simple NMSSM with a part of invariant super-potential when $\mu = \mu' = \xi_F = 0$

$$W_{NMSSM} = \lambda\widehat{S}\widehat{H}_u.\widehat{H}_d + \frac{\kappa}{3}\widehat{S}^3. \tag{4}$$

Then, eliminating the parameters $m_3^2, m_s^{'2}$ and ξ_S in (3), combining the vevs of $\hat{S}$ in the weak sector or in the breaking supersymmetry to define μ:

$$\mu_{eff} = \lambda s. \tag{5}$$

The matter of μ in MSSM has been solved then.
As any supersymmetry theory with invariant super-potential sector (ternary), the Lagrangians, which contain the soft supersymmetry violation conditions specified by (3), have one Z_3 symmetry randomly, which is corresponding to the multiplication of all chiral superfields with $e^{2\pi i/3}$. The invariant super-potential (4) is presented like the invariant Z_3. The symmetry Z_3 will be broken by non-dimension terms in the super-potential (1). The model with super-potential (1) is the NMSSM. The invariant Z_3 Higgs sector is defined by seven parameters

Journal of Physics: Conference Series **537** (2014) 012015 doi:10.1088/1742-6596/537/1/012015

$\lambda, \kappa, m_{H_d}^2, m_{H_u}^2, m_S^2, A_\lambda, A_\kappa$. The expressions of Higgs mass matrix in the invariant Z_3 of the NMSSM show that invariant Z_3 is obtained when:

$$m_3^2 = m_S^{2\prime} = \xi_S = \mu = \mu' = \xi_F = 0. \tag{6}$$

From the supersymmetry gauge interaction and soft supersymmetry breaking conditions, we obtain Higgs potential:

$$\begin{aligned} V_{Higgs} = &\left|\lambda(H_u^+H_d^- - H_u^0H_d^0) + \kappa S^2 + 2\mu' S + \xi_F\right|^2 \\ &+(m_{H_u}^2 + |\mu + \lambda S|^2(\left|H_u^0\right|^2 + \left|H_u^+\right|^2) + (m_{H_d}^2 + |\mu + \lambda S|^2(\left|H_d^0\right|^2 + \left|H_d^-\right|^2) \\ &+\tfrac{g_1^2+g_2^2}{8}(\left|H_u^0\right|^2 + \left|H_u^+\right|^2 - \left|H_d^0\right|^2 - \left|H_d^-\right|^2)^2 + \tfrac{g_2^2}{2}\left|H_u^+H_d^{0*} + H_u^0H_d^{0*}\right|^2 \\ &+m_S^2|S|^2 + (\lambda A_\lambda(H_u^+H_d^- - H_u^0H_d^0)S + \tfrac{1}{3}\kappa A_\kappa S^3 + m_3^2(H_u^+H_d^- - H_u^0H_d^0) \\ &+m_s'^2S^2 + \xi_S + h.c. \end{aligned} \tag{7}$$

where g1 and g2 present gauge interaction U(1) and SU(2) .
The even-CP Higgs sector has three mixed states, which are the real parts of H_u, H_d and S. The largest mass state h, which is nearly the same with SM, interacts with electroweak Gauge boson, this state has squared mass M_h^2 given by [5, 6]:

$$M_h^2 = M_z^2\cos^2 2\beta + \lambda^2 v^2 \sin^2 2\beta + rad.corrs + \Delta_{mix} \tag{8}$$

The diagonal elements of matrix in the singlet sector are given by (with the assumption that s $> v_u, v_d$)

$$M_{ss}^2 \simeq ks\,(A_k + 4ks) \tag{9}$$

The term Δ_{mix} in (8) is derived from the mixture of doublet and singlet

$$\Delta_{mix} \simeq \frac{4\lambda^2 s^2 v^2(\lambda - k\sin 2\beta)^2}{\overline{M}_h^2 - M_{SS}^2} \tag{10}$$

where $\overline{M}_h^2$ is given by M_h^2 and has no mixed term.
Studying the decay of new particles in the Model will bring us the hope of finding out these particles as well as verifying the correctness of that Model [8]. In this paper we study some decays of neutral Higgs bosons. The numerical results are also shown in graph to evaluate the influence of m_{a_1} and ϕ on the decay width and on the lifetime of Higgs bosons.

2. Some decays of neutral Higgs bosons in the NMSSM

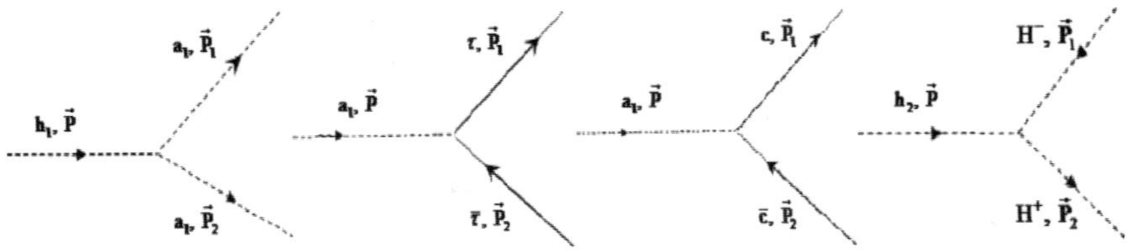

Figure 1. Feynman diagram for some decays of neutral Higgs.

The amplitudes which are calculated to tree level have the following forms:

$$M_0(h_1 \to a_1a_1) = \frac{gm_z}{2cos\theta_w}\cos\beta.cos\,(\alpha+\beta) \tag{11}$$

$$M_0(a_1 \to \tau\overline{\tau}) = \overline{u}^{\alpha}(\overrightarrow{P_1}, s)\left(\frac{igm_\tau}{2m_{\rm w}}\right)\tan\beta\,(\gamma_5)^{\beta}_{\alpha}\,\nu_\beta(\overrightarrow{P_2}, s) \tag{12}$$

$$M_0(a_1 \to c\bar{c}) = \bar{u}^{\alpha}\left(\vec{P}_1, s\right)\left(\frac{gm_c}{2m_{\rm w}\sin\beta}U^P_{12}\,(\gamma_5)^{\beta}_{\alpha}\right)\upsilon_\beta\left(\vec{P}_2, s\right) \tag{13}$$

$$M_0(h_2 \to H^+H^-) = A = -igm_w\left(U^S_{a1}\cos\beta + U^S_{a2}\sin\beta\right)$$

$$-\frac{igm_w}{2\cos\theta_w}\left(U^S_{a1}\sin\beta + U^s_{a1}\cos\beta\right).cos2\beta + i\frac{\lambda^2}{\sqrt{2}}\left(v_1U^S_{a2} + v_2U^S_{a1}\right)\sin 2\beta$$

$$-\frac{i}{\sqrt{2}}\lambda U^S_{a3}\left[(2kx + A_\lambda)\sin\beta + 2\lambda x\right] \tag{14}$$

The decay widths which are calculated to tree level are:

$$\Gamma_0(h_1 \to a_1a_1) = \frac{g^2m_z^2}{64\pi cos^2\theta_{\rm w}}\frac{\cos^2\beta cos^2(\alpha+\beta)\left(m_h^2 - 4m_{a1}^2\right)^{1/2}}{m_h^2} \tag{15}$$

$$\Gamma_0(a_1 \to \tau\overline{\tau}) = \frac{g^2m_\tau^2}{64\pi m_{\rm w}^2m_{a_1}^2}\tan^2\beta\left(2m_{a_1}^2 - 3m_\tau^2\right)\left(m_{a_1}^2 - 4m_\tau^2\right)^{1/2} \tag{16}$$

$$\Gamma_0(a_1 \to c\bar{c}) = \frac{\left|U^P_{12}\right|^2 g^2m_c^2(m_{a_1}^2 - 2m_c^2)}{128\pi m_{a_1}^3m_{\rm w}^2\sin^2\beta}\sqrt{m_{a_1}^4 - 4m_{a_1}^2m_c^2} \tag{17}$$

$$\Gamma_0(h_2 \to H^+H^-) = \frac{|A|^2}{16\pi m_{h2}^3}\sqrt{m_{h2}^4 - 4m_{h2}^2m_H^2} \tag{18}$$

3. Numerical Results

To study the influence of the mass m_{a1} and ϕ on the Higgs decay process, we use two set of parameters [5, 9] for programming numerical calculation: λ = 0,8; x = $200e^{i\phi}$; k = 0,1; m_H = 100; tanβ= 3; sin α = - 0,58; A_k = 6; A_λ = 486. Perform the numerical calculations we obtain the graphs from Fig.2 to Fig.6.

From Fig.2, when considering the variation of m_{a_1} from $4GeV < m_{a_1} < 8GeV$, we can see that the variation can decrease the value of decay width by 0,05% and increase the value of the lifetime of h_1 to 0,05% . The obtained result of decay width $h_1 \to a_1a_1$ is around 0,00633 - 0,00636(1/s), and the lifetime of h_1 is around 157,2 - 158,0 (s)

Fig.3 shows that the influence of m_{a_1} on the decay width and on the lifetime of a_1 in the decay $a_1 \to \tau\overline{\tau}$ is relatively significant. When considering the variation $4GeV < m_{a_1} < 8GeV$, we see that the variation can increase the value of decay width to 5 times and decrease the value of the lifetime of a_1 by 5 times. The obtained result of decay width $a_1 \to \tau\overline{\tau}$ is around 0,000015 - 0,00006(1/s), and the lifetime of a_1 is around 157,2 - 158,0 (s).

Fig.4 shows that the influence of m_{a_1} on the decay width and on the lifetime of a_1 in the decay $a_1 \to c\bar{c}$ is also relatively significant. When considering the variation $4GeV < m_{a_1} < 8GeV$, we see that the variation can increase the value of decay width to 3 times and decrease the

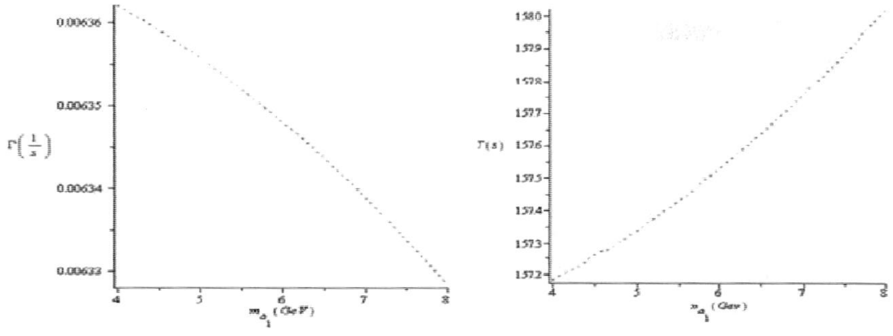

Figure 2. Influence of m_{a_1} on the decay width and on the lifetime of h_1 in the decay $h_1 \to a_1 a_1$.

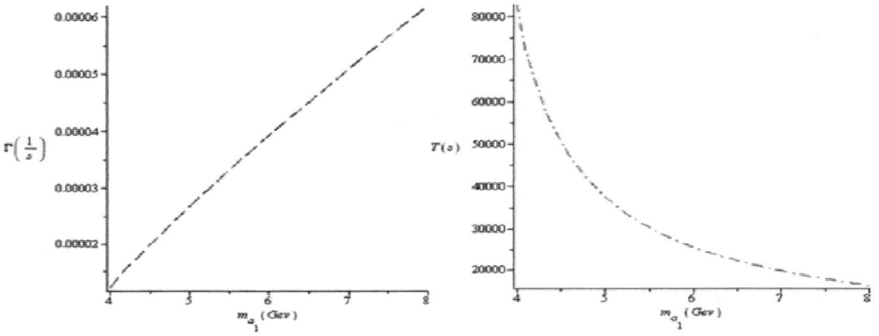

Figure 3. Influence of m_{a_1} on the decay width and the lifetime of a_1 in the decay $a_1 \to \tau\bar{\tau}$.

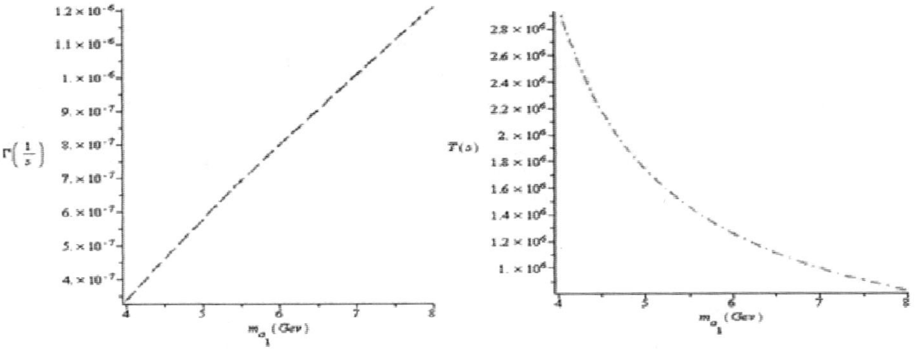

Figure 4. Influence of m_{a_1} on the decay width and on the lifetime of a_1 in the decay $a_1 \to c\bar{c}$.

value of the lifetime of a_1 by 3 times. The obtained result of decay width $a_1 \to c\bar{c}$ is around 4.10^{-7}–12.10^{-7} (1/s), and the lifetime of a1 is around $0,9.10^6$–$2,9.10^7$ (s).
If the CP violation is considered, the influence of ϕ (choose $m_{a_1} = 5$GeV) on the decay width and the lifetime of a_1 in the decay $a_1 \to c\bar{c}$ is quite significant (Fig.5).
Specifically, the influence of ϕ on the decay width and on the lifetime of a_1 in the decay $a_1 \to c\bar{c}$ is relatively significant. If we consider the variation from $0 < \phi < 0.2(rad)$, the variation can increase value of decay width to 0,6% and decrease the value of the lifetime of a_1 by 2% .The obtained result of decay width $a_1 \to c\bar{c}$ is about $5,75.10^{-7} - 6,3.10^{-7}$ (1/s), and the lifetime of a_1 is about $1,59.10^6 - 1,73.10^6$ (s).

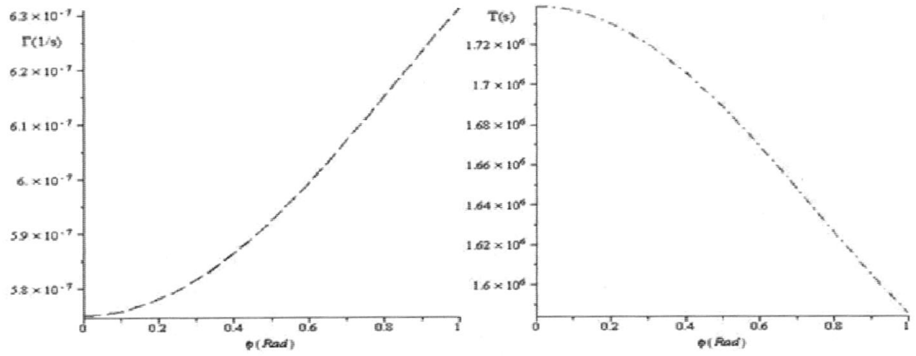

Figure 5. Influence of ϕ on the decay width and on the lifetime of a_1 in the decay $a_1 \to c\bar{c}$.

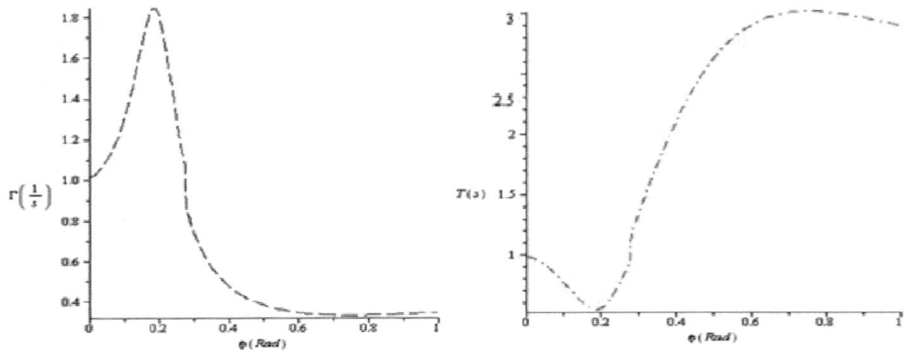

Figure 6. Influence of ϕ on the decay width and on the lifetime of a_1 of the decay $h_2 \to H^+H^-$.

From Fig.6 we can see that the influence of ϕ on the decay width and on the lifetime of h_2 in the decay $h_2 \to H^+H^-$ is relatively significant. If we consider the variation from $0 < \phi < 0.2(rad)$, the variation can contribute to increase the value of decay width to 60% and decrease the value of the lifetime of h_2 by 40%. The obtained result of decay width $h_2 \to H^+H^-$ is about 0,4 - 1,8(1/s), and the lifetime of h_2 is about 0,55 - 3(s).

4. Conclusion

In the NMSSM, a single superfield is added with complex scalar field components, this leads to the appearance of seven Higgs in the NMSSM (including three even-CP Higgs $\mathrm{h}_{1,2,3}$ ($\mathrm{m}_{\mathrm{h}1} < \mathrm{m}_{\mathrm{h}2} < \mathrm{m}_{\mathrm{h}3}$), two odd-CP Higgs $\mathrm{a}_{1,2}$ ($\mathrm{m}_{\mathrm{a}1} < \mathrm{m}_{\mathrm{a}2}$) and a pair of charged Higgs $H^{\pm}$). We have obtained many interesting results from studying some decays of neutral Higgs in this paper:

- There are many decay channels of neutral Higgs in the NMSSM, especially the decays Higgs to Higgs
- The influence of $\mathrm{m}_{\mathrm{a}1}$ and ϕ on some decays is relatively significant. We need to pay attention to this element when studying theories as well as paying attention to the decay experiments of neutral Higgs
- The lifetime of neutral Higgs in some decays is relatively short, and this brings us the hope that we can find Higgs soon.

Acknowledgment
This research is supported by the National Foundation for Science and Technology Development (NAFOSTED) of Vietnam. Grant number: 103.03-2012.80.

References
[1] Radovan Demisenk and John F. Gunion, Phys. Rev. D79 (2009) 055014.
[2] M.M. Almarashi anh S.moretti, Phys. Rev. D85 (2012) 017701.
[3] M.M. Almarashi anh S.moretti, Phys. Rev. D84 (2011) 015014.
[4] H. E. Haber and G.L. Kane. Phys. Rep. 117 (1985) 75.
[5] Ulrich Ellwanger, hep-ph/1108.0157.
[6] U. Ellwanger, C. Hugonie and A. M. Teixeira, Phys. Rept. 496 (2010)1.
[7] W. Bernrenther and M. Suzuki, Rev. Mod. Phys. 63 (1991) 3-13.
[8] H.H.Bang, D.T.L.Thuy and N.C.Cuong, Commu. in Phys. Vol.14 (2004) 157-164.
[9] Radovan Dermisek (2010), hep-ph/1012.3487vl.
[10] A. Barlt, et. al., Phys. Lett. B419 (1998) 243.

IWTCP1 & NCTP38 IOP Publishing
Journal of Physics: Conference Series **537** (2014) 012016 doi:10.1088/1742-6596/537/1/012016

One-Loop β Functions for Yukawa Couplings in the Electroweak-Scale Right-Handed Neutrino Model

Nguyen Nhu Le[1] **and Pham Quang Hung**[2]
[1]Physics department, Hue University's College of Education, Vietnam
[2]Department of Physics, University of Virginia, USA

E-mail: naspears13@gmail.com

Abstract. Fermions in the model of electroweak-scale right-handed neutrinos (EWRH) with masses of the order of 300 GeV or more could result in dynamical electroweak symmetry breaking by forming condensates through the exchange of a fundamental Higgs scalar doublet or triplet. These condensates are dynamically studied within the framework of the Schwinger-Dyson equation. With the electroweak symmetry broken by condensates, the fully worked-out model of EWRH in which there are two doublets and two triplets, one of which is composite and the others being the original fundamental scalar doublet and triplet could be suitable for recent LHC discovery of the 125 GeV scalar particle.

1. Introduction

The SM has been established for a long time and has shown its abilities to address many problems in High Energy Physics. Nevertherless, some problems can not be explained by SM. One of which is SM Higgs mechanism where spontaneous symmetry breaking (SSB) is put in by hand. With Higgs potential $V(\phi) = \mu^2\phi^+\phi + \lambda(\phi^+\phi)^2$, this mechanism does not explain why μ^2 is negative when it could, a priori, have been positive. μ^2 and hence $m_H^2 = -2\mu^2 = 2\lambda v^2$ are unstable to large radiative corrections from much higher enery scales (gauge hierarchy problem, fine-tuning needed to keep Higgs light, etc.). Moreover, the SM Yukawa mechanism for generating fermion masses, with $m_f = y_f v/\sqrt{2}$, does not give insight into these masses, since it just puts in the value of Yukawa couplings y_f by hand, and some y_f's range down to 10^{-5} with no explanation. Might it be a possible way to give a dymanical reason for fermion masses? Consider two major previous cases of superconductivity and Gell-Mann Levy σ model for spontaneous chiral symmetry breaking (SχSB) where the fundamental scalar fields were used to model SSB. For these cases, the actual underlying physics did not involve fundamental scalar fields but instead a bilinear fermion condensate. For instance, the physical properties of superconductor materials can be explained by dynamically forming a condensate state of Cooper pairs (ee) in BCS theory; since these are charged, they give mass to photon, resulting in Meissener effect. In the second case, the actual origin of SχSB in QCD is the dynamical formation of a $(\bar{q}q)$ condensate. These previous models of symmetry breaking, therefore, bring about the underlying physics responsible for this symmetry breaking may well be a dynamically induced fermion condensate.
The condition for condensation mentioned in [1] is that the Yukawa couplings are larger than critical Yukawa couplings α^c. For instance, when the energy scale is of the order of $O(1\text{TeV})$, the Yukawa coupling of a heavy fourth generation in [2] is large enough, exceeding α^c, to form

Journal of Physics: Conference Series **537** (2014) 012016 doi:10.1088/1742-6596/537/1/012016

condensates through the exchange of a fundamental Higgs scalar doublet. In this paper, a similar mechanism has been discussed. We have also investigated the one-loop β functions in the model of [3] to find the energy scale forming condensate states of fermions. For the fact that condensates of mirror fermions can be a part of DEWSB, the fully worked-out model of EWν_R in which there are one singlet, two doublets and two triplets could be compatible with recent LHC discovery of Higgs particle [4], [5], while the SM with fourth generations has been ruled out.

The plan of the paper is as follows. Firstly, we summarize the essential elements of the EWν_R model of [3]. Secondly, the analytical formula and numerical calculations related to the one-loop beta functions of Yukawa coulings are presented. We conclude with some remarks concerning the DEWSB in the model of [3].

2. The EWν_R model

The essential elements of the model presented in [3] can be reviewed as below

The $SU(2)_L \times U(1)_Y$ fermion content of the EWν_R model of [3]

- $SU(2)_L$ lepton doublets:

$$\begin{aligned} SM : l_L &= \begin{pmatrix} \nu_L \\ e_L \end{pmatrix}; \\ Mirror : l_R^M &= \begin{pmatrix} \nu_R \\ e_R^M \end{pmatrix}, \end{aligned} \tag{1}$$

for the SM left-handed lepton doublet and for the right-handed mirror lepton doublet respectively.

- $SU(2)_L$ lepton singlets:

$$SM : e_R; \tag{2}$$

$$Mirror : e_L^M \tag{3}$$

for the SM right-handed lepton singlet left-handed mirror lepton singlet respectively.

Similarly, for the quarks, we have

- $SU(2)_L$ quark doublets:

$$\begin{aligned} SM : q_L &= \begin{pmatrix} u_L \\ d_L \end{pmatrix}; \\ Mirror : q_R^M &= \begin{pmatrix} u_R \\ d_R^M \end{pmatrix}, \end{aligned} \tag{4}$$

for the SM left-handed quark doublet and for the right-handed mirror quark doublet respectively.

- $SU(2)_L$ quark singlets:

$$SM : u_R, d_R; Mirror : u_L^M, d_L^M, \tag{5}$$

for the SM right-handed lepton singlet left-handed mirror lepton singlet respectively.

Higgs content in the model of EWν_R

-Higgs doublets

$$\Phi = \begin{pmatrix} \phi^+ \\ \phi^0 \end{pmatrix} = (1, 2, Y/2 = 1/2). \tag{6}$$

-Higgs triplets

$$\widetilde{\chi} = \frac{1}{\sqrt{2}}\vec{\tau} \cdot \vec{\chi} = \begin{pmatrix} \frac{1}{\sqrt{2}}\chi^+ & \chi^{++} \\ \chi^0 & -\frac{1}{\sqrt{2}}\chi^+ \end{pmatrix} = (1, 3, Y/2 = 1), \tag{7}$$

$$\xi = \begin{pmatrix} \xi^+ \\ \xi^0 \\ \xi^- \end{pmatrix} = (1, 3, Y/2 = 0)\,. \tag{8}$$

-Higgs singlet

$$\phi_S = (1, 1, Y/2 = 0). \tag{9}$$

Higgs triplets such as in (7) were considered earlier in various contexts [6],[7],[8].
In the search for the beta functions discussed in this work, it is important to know Yukawa Lagrangian. The Yukawa couplings can be found in [9]. In the case of SM fermions, we have the usual Yukawa interactions

$$\mathcal{L}_{SM} = -g_{SM}\bar{\Psi}_L\Phi\Psi_R + \text{h.c.}. \tag{10}$$

For mirror fermions we consider the terms

$$\mathcal{L}_{e^M} = -g_l^M \bar{l}_R^M \Phi e_L^M + \text{h.c.}, \tag{11}$$

$$\mathcal{L}_{d^M} = -g_d^M \bar{q}_R^M \Phi d_L - g_d^M \bar{q}_R^M \widetilde{\Phi} d_L^M + \text{h.c.}, \tag{12}$$

$$\mathcal{L}_{\nu_R} = g_M l_R^{M,T} \sigma_2 \tau_2 \widetilde{\chi} l_R^M, \tag{13}$$

$$\mathcal{L}_S = -g_{Sl}\bar{l}_L l_R^M \phi_S + \text{h.c.}. \tag{14}$$

As energy increases, the Yukawa interactions become strong, bound states of fermions can be formed. The top quark, however, can not form a bound state by exchanging the Higgs boson [2] because of a severely fine-tuned picture of DEWSB. Moreover, the vacuum expectation values of Higgs doublets and Higgs triplets break the global $SU(2)_L \times SU(2)_R$ down to the custodial $SU(2)_D$. Thus, we are interested in the running of Yukawa couplings: g_l^M, g_d^M and g_M.

3. Oneloop beta functions for Yukawa couplings in the EWν_R model

3.1. Analytical formula

In the EWν_R model, right-handed neutrino interacts with Higgs triplet $\widetilde{\chi}$. Hence, the running of Yukawa coupling g_M is firstly investigated in this section. The one loop contributions to the Yukawa coupling renormalization constant are composed of four terms: vertex corrections, fermion self-energy, scalar self-energy and gauge interactions. At high enery scale, gauge couplings, however, become too small compared to Yukawa couplings then gauge contributions to the one-loop β functions can be neglected. For this reason the first three terms are computed. Consider the vertex corrections which contribution diagrams modifying the proper one-loop β function for g_M are shown in fig. 1

$$\Gamma_1 = -(i)^3 g_M^3 \left(1 + \frac{1}{2}\right) \int \frac{d^4k}{(2\pi)^4} \left(\frac{i}{\gamma^\mu p_\mu - \gamma^\mu k_\mu}\right)^2 \frac{i}{k^2}. \tag{15}$$

The contribution coming from equation (15) is computed by using dimensional regularization. We have

$$\beta_1 = 3 \times \frac{g_M^3}{16\pi^2}. \tag{16}$$

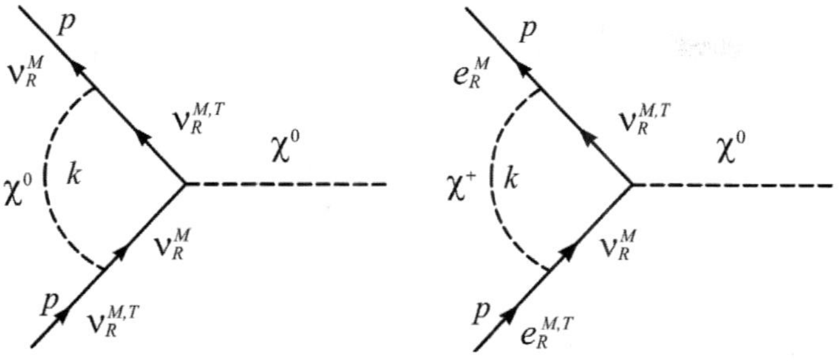

Figure 1. Contribution diagrams of vertex corrections modifying the proper one-loop β function for g_M

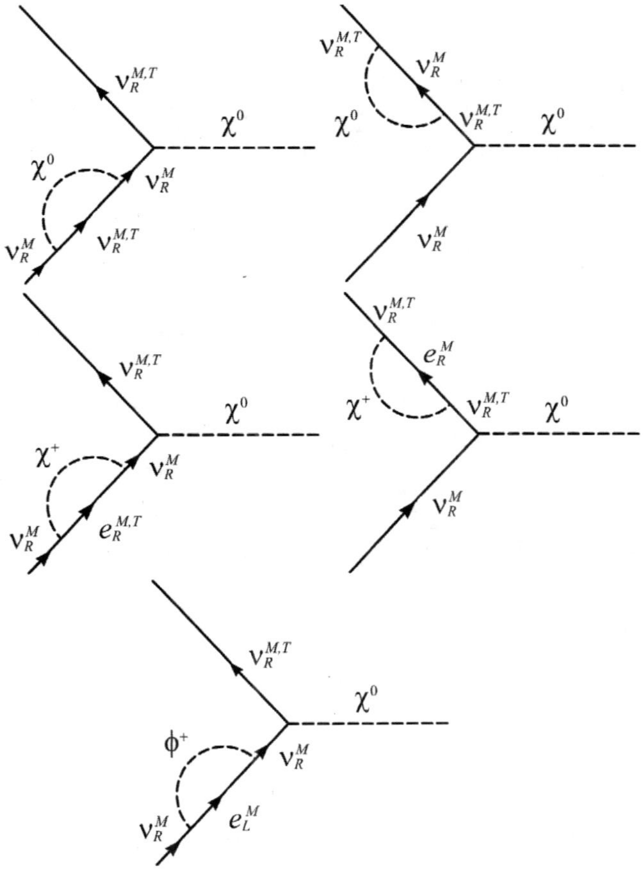

Figure 2. Contribution diagrams of femion self-energy modifying the proper one-loop β function for g_M

The fermion self-energy term, which diagrams shown in fig. 2, is given by

$$\Sigma_f(p) = \left[2(-ig_M)^2 + 2\left(-\frac{i}{\sqrt{2}}g_M\right)^2 + \left(-ig_l^M\right)^2\right] \int \frac{d^4k}{(2\pi)^4} \frac{i}{k^2} \frac{i(\gamma^\mu p_\mu \gamma^\mu k_\mu)}{(p-k)^2}, \quad (17)$$

and its contribution to the beta function is

$$\beta_2 = \left(\frac{3}{2}g_M^2 + \frac{1}{2}g_l^{M2}\right)\frac{g_M}{16\pi^2}. \tag{18}$$

Finally, the Feynman diagram for the scalar self-energy is shown in fig. 3 and

$$\Sigma_s(p) = (-ig_M)^2(-)\int\frac{d^4k}{(2\pi)^4}Tr\left(\frac{i}{\gamma^\mu k_\mu}\frac{i}{\gamma^\mu p_\mu - \gamma^\mu k_\mu}\right), \tag{19}$$

$$\beta_3 = 2\times\frac{g_M^3}{16\pi^2}. \tag{20}$$

The total contribution to the beta function is then

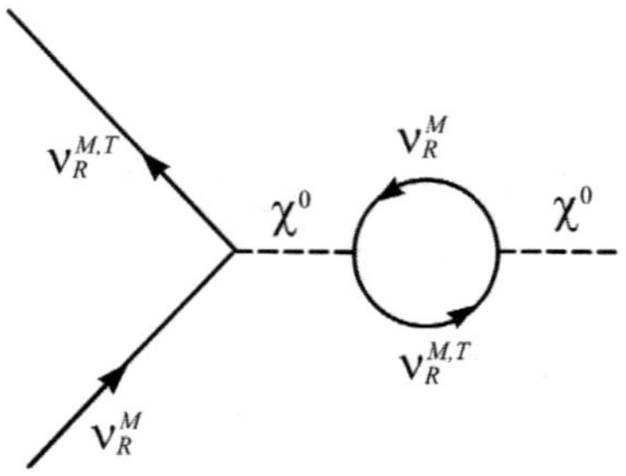

Figure 3. Contribution diagram of scalar self-energy modifying the proper one-loop β function for g_M

$$\beta_{g_M} = \frac{dg_M}{dt} = \beta_1 + \beta_2 + \beta_3 = \frac{\left(13g_M^3 + g_L^{M2}g_M\right)}{32\pi^2}. \tag{21}$$

Similarly, by using dimensional regularization, we found the the evolution of Yukawa couplings of both mirror leptons and mirror quarks

$$\beta_{g_l^M} = \frac{dg_l^M}{dt} = \frac{22g_l^{M3} + 3g_M^2 g_l^M}{64\pi^2}, \tag{22}$$

$$\beta_{g_d^M} = \frac{dg_d^M}{dt} = \frac{3g_d^{M3}}{8\pi^2}. \tag{23}$$

3.2. Numerical calculations

As shown in fig.4, with initial values of g_M, g_l^M, and g_d^M being $2.50, 2.46$ and 2.73 respectively or the coresponding induced masses being about 300 GeV, the Yukawa couplings of mirror fermions increase dramatically as energy increases and Ladau pole singularrities appear at the order of $O(1\text{TeV})$. Moreover, by using the Schwinger-Dyson equation in the rainbow approximation to compute the self-energy of fermions and subsequently their condensates, we found the critical Yukawa couplings of g_M, g_l^M, and g_d^M are π, π and $\pi/2$ respectively [10]. Above which condensates carrying the electroweak quantum numbers can be formed which could spontaneously break electroweak symmetry. It is amazing that from fig.4 both mirror fermions form condensates at the order of $O(1\text{TeV})$ while top quark which initial Yukawa coupling is approximately 1 hardly to do so. For this reason, mirror fermions are responsible for DEWSB in the EWν_R model.

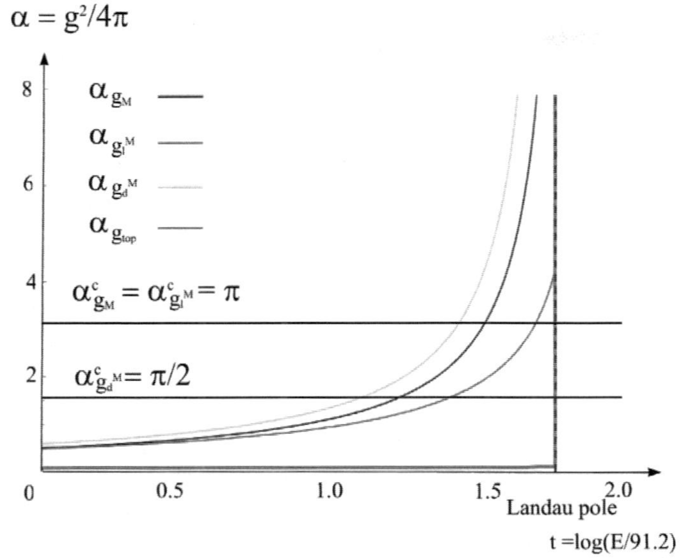

Figure 4. The evolution of Yukawa couplings

4. Conclusion

The central results of our paper, on one hand, is eqs.(21), (22) and (23) which give beta functions for the Yukawa couplings in the EWν_R model. For suitble initial values of Yukawa couplings, the Landau singularities appear at the order of $O(1\text{TeV})$. As critical Yukawa couplings are known, electroweak symmetry can be broken dynamically by condensates of mirror quarks and right-handed neutrinos. On the orther hand, because DEWSB scale is of the order of O(1TeV) with forbidden case that mirror leptons get condensate states later than the others do (badly symmetry broken), the mass distribution of fermions in the EWν_R model and constraints on form factors by using conditions assuring proper alignment of the two VEVs can be investigated in near future.

Acknowledgment

The authors would like to thank Professors in Department of Physics, University of Virginia for inspiring discussions at earlier stages of the present study. Furthermore, we are indebted to Dr. Vo Tinh for critical remarks. Financial support of Hue University is gratefully acknowledged.

References

[1] C. N. Leung, S. T. Love, *Nucl. Phys.*, **B273** (1986) 649.
[2] P. Q. Hung, C. Xiong, *Nucl. Phys.*, **B848** (2011) 288.
[3] P. Q. Hung, *Phys. Lett. B*, **649**, (2007) 275.
[4] S. Chatrchyan *et all.*, [CMS Collaboration], *Phys. Lett.* **B716** (2012) 30.
[5] S. Chatrchyan *et all.*, [CMS Collaboration], *Phys. Lett.* **B716** (2012) 30.
[6] H. Georgi and M. Machacek, *Nucl. Phys.*, **B262** (1985) 463.
[7] R. S. Chivukula and H. Georgi, *Phys. Lett. B.*, **182** (1986) 181.
[8] P. H. Frampton, M. C. Oh, and T. Yoshikawa, *Phys. Rev.*, **D66** (2002) 033007.
[9] P. Q. Hung, *Phys. Lett. B.*, **659** (2008) 585.
[10] P. Q. Hung and N. N. Le, In perparation.

AUTHOR INDEX

Institute of Physics
Dirac House, Temple Back
Bristol BS1 6BE UK

CURRAN ASSOCIATES INC
proceedings.com